AUTOMOTIVE DIESELS

EDWARD RALBOVSKY

DELMAR PUBLISHERS INC.

Delmar Staff
  Administrative Editor: Mark W. Huth
  Production Editor: Ann Drylewski

For information, address Delmar Publishers Inc.
2 Computer Drive, West, Box 15-015
Albany, New York 12212

Printed in the United States of America
Published simultaneously in Canada
by Nelson Canada,
a division of International Thomson Limited

10 9 8 7 6 5 4 3 2

Library of Congress Cataloging in Publication Data

Ralbovsky, Edward.
  Automotive diesels.
  Includes index.
  1. Automobiles—Motors (Diesel) 2. Automobiles—
Motors (Diesel)—Maintenance and repair. I. Title.
TL229.D5R35   1985   629.2′506   84-19956
ISBN 0-8273-2217-8
ISBN 0-8273-2218-6 (instructor's guide)

CONTENTS

NOTICE TO THE READER

PREFACE

The diesel engine, a very popular source of power for large trucks, tractors, and construction equipment, is now a viable option in automobiles and light trucks. On a wide variety of makes and models, the consumer can opt to buy a diesel-powered vehicle. Excellent fuel economy, a major advantage of the diesel, may fit today's buyers' needs.

Naturally, as more diesel-powered cars and trucks enter the national fleet, more automotive technicians will be required to service them. Much of the knowledge already gained through gasoline engine experience will apply, but there are many differences that will require specialized knowledge.

There are systems and components unique to the diesel engine. Also, the diesel engine used in cars and light trucks is different from diesel engines used in other applications. Though the fundamentals of operation are the same, the automobile diesel has different performance standards and criteria from, for example, a tractor diesel. This means design and related systems will be different, requiring different service procedures and diagnostic techniques. The purpose of this textbook is to point out the differences and discuss systems and procedures unique to diesel engines used in cars and light trucks.

To accomplish this task, the text begins with the fundamentals of diesel engine operation and why the automobile diesel is different. Contrasts are often made with the gasoline engine. This textbook assumes that the reader will already know gasoline engine fundamentals and operation. This belief is founded on the premise that automotive programs will begin with the more familiar and popular gasoline engine, then go on to the diesel engine. Because of this premise, the reader will not find chapters on hand tool use, measuring, or using the metric system, for example. It is assumed the reader has had that experience.

Chapters 5 through 9 discuss diesel fuel systems in great detail. With the exception of the engine, it is the most expensive and certainly the most complex. Accurate, consistant diagnosis can only be accomplished when the technician has a thorough background on what is happening within the system and how related components affect each other. It is imperative that the cautions are strictly adhered to and that the manufactuer's service manual be used. This is critical for specifications change and each manufacturer has a limit on how much service can be performed on the fuel system.

Chapters 10 and 11 discuss the diesel's starting systems and emission controls. Many similarities to gasoline engines will be noted by the reader. Also, new and different systems such as the glow plug system are discussed.

Chapters 12 through 17 discuss the diesel engine itself. It's basic format is that the reader is disassembling then assembling the engine. These chapters are put toward the back of the book for some basic reasons. First, if the diesel engine is to be disassembled, the fuel system and glow plug system must be disconnected then reconnected properly. A knowledge of how they work and advance warning of the cautions involved could save much grief later on. Second, it is far more likely that the fuel, starting, and emission systems are the cause of problems rather than the engine itself. Proper diagnosis of these systems could prevent unwarranted disassembly of the engine. Furthermore, they may be the cause of an engine failure.

The purpose of chapters 12 through 17 is to point out the features and procedures different from the comparable gasoline engine. These chapters do not discuss operations common to both engines such as grinding valves, honing cylinders, and so on. It is assumed the reader has had experience in performing these operations and it would be redundant to include it in this text.

Chapter 18 contains normal maintenance procedures and a diagnostic guide listing symptoms, causes, and repairs.

The service procedures in this text are intended to be only a guide, to explain what is important in the procedure and why it's important. A service manual should always be consulted whenever performing service on the vehicle because it will have the specific procedures and specifications for the particular make, model, and year of the vehicle being worked on.

When finished with the text, the student/technician will know the fundamentals, operation, and service of diesel engines used in cars and light trucks. With this knowledge of diesel engines, the students/technicians can increase their employability, value to themselves and the place they work.

ACKNOWLEDGMENTS

The manuscript for AUTOMOTIVE DIESELS was thoroughly reviewed by several experts and diesel instructors. The author and the publisher especially thank Gerald J. Bowers, Milwaukee Area Technical College; Richard Radock, Westmoreland Community College; Tom Hogue, Santa Ana College; Francis Hall, State University of New York at Canton; Fran Treichler, Ford Motor Company. Appreciation is also expressed to the following persons and organizations:

Robert Bosch Corporation
Chevrolet Division
Cummins Engine Company, Inc.
John Deere & Company
Exxon Inc.
Ford Motor Company
General Motors Product Service Training
Mace Motors Inc.
Mercedes-Benz of North America Inc.
Oldsmobile Division
Stanadyne Diesel Systems
T.J. Toyota, Inc.
Toyota Motor Sales, Inc.
United Technologies Diesel Systems
Volkswagen of America, Inc.
All other photographs courtesy of Mark Scanlin

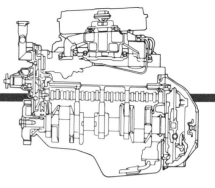

chapter 1
Introduction

Objectives

In this chapter you will learn:

- How diesel engines were developed
- Diesel engine applications
- Why the passenger car diesel is different from other diesel engines
- The characteristics of the diesel engine (advantages and disadvantages when compared to the gasoline engine)

HISTORY

In the late nineteenth century, research was conducted to develop more efficient engines that would run on energy sources other than steam. One of the pioneers was Dr. N.A. Otto. Dr. Otto succeeded in making a gasoline engine that ran on a cycle he had discovered in 1876. The *Otto cycle* is the basis on which gasoline and diesel engines operate today.

Once the Otto cycle was established, researchers began searching for methods to satisfactorily ignite a mixture of air and fuel. One method was to compress the air in a

cylinder so tightly that extremely high temperatures would be produced. Fuel would be injected into this very hot air and would burn almost immediately. The expansion of the burning air and fuel produced power, figure 1-1. This type of engine is called the *compression-ignition engine.* However, several problems arose. The principal ones were a lack of building materials that could hold up under high temperatures, and difficulties in injecting fuel into a high-pressure area.

Credit for developing the compression-ignition engine is given to a German engineer named Rudolph Diesel. Initially, he attempted an engine that ran on coal dust but these engines exploded. In 1894, he successfully switched to a liquid fuel. His engine was more economical and efficient than the known engines of his day, figure 1-2. Finally, in 1895, Diesel was granted the U.S. patents on the compression-ignition engine.

By 1900, the diesel engine was widely used in industrial plants in Europe. These diesel engines, being large and heavy, were used only in stationary applications.

As diesel engine technology matured, the power-to-weight ratio increased. By 1924 the first diesel-powered ocean liner was launched; in 1925 a diesel-powered bus;

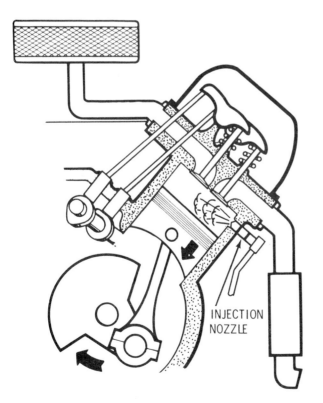

Figure 1-1 The burning air/fuel mixture expands pushing the piston downward. (*Courtesy of Oldsmobile Division*)

Figure 1-2 Rudolph Diesel's first successful engine (*Courtesy of Ford Motor Company*)

1

and in 1929 a diesel-powered truck, figure 1-3. These engines were still too large and bulky for automobile use, although Peugeot attempted the feat in 1922.

In 1927, the Robert Bosch Company began making fuel injection equipment for diesels. This was important because Robert Bosch was able to mass-produce fuel injection equipment that allowed diesel use in a variety of applications. Furthermore, his company granted licenses to manufacturers in other countries. This created greater production facilities and allowed the diesel engine to become a worldwide power source.

By the 1930s, diesel engines were widespread. In 1936, Mercedes-Benz began producing a diesel-powered passenger car, figure 1-4.

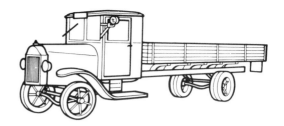

Figure I-3 The first diesel-powered vehicle (*Courtesy of Ford Motor Company*)

TYP		ENGINE	1.PRODUC-TION YEAR	DISPLACE-MENT cm³	STROKE mm	BORE mm	MAX OUTPUT AT SPEED (DIN) kW	1/min	TORQUE AT SPEED Nm	1/min	NUMBER OF CYLINDERS	DRY ENGINE WEIGHT KG
	260D	OM138	1936	2545	100	90	33	3300			4	
	170D	OM636	1949	1697	100	73,5	28	3200	96	2000	4	
	180D	OM636	1953	1767	100	75	31	3500	101	2000	4	
	190D	OM621	1958	1897	83,6	85	37	4000	108	2200	4	
	200D	OM621	1965	1988	83,6	87	40	4200	113	2400	4	
	200D	OM615	1967	1988	83,6	87	40	4200	113	2400	4	
	220D	OM615	1967	2197	92,4	87	44	4200	126	2400	4	
	240D	OM616	1973	2404	92,4	91	48	4200	137	2400	4	
	300D	OM617	1974	3005	92,4	91	59	4000	172	2400	5	
	200D	OM615	1976									195
	220D	OM615	1976									197
	240D	OM616	1976		AS ABOVE							197
	300D	OM617	1976									229
	300CD	OM617	1977									229
	300SD	OM 617 A	1978	2998	92,4	90,9	85	4200	235	2400	5 TURBO CHARGED	244

Figure I-4 Note on the chart that engine output and torque increased from the 260D to the 300SD models. (*Courtesy of Mercedes-Benz of North America Inc.*)

After World War II, gasoline was expensive in markets other than the United States, forcing Europe and Japan to develop economical diesel engines for passenger cars and small trucks. In the United States, diesel engine development advanced in the area of construction and large truck equipment, figure 1-5. Car buyers in the United States chose power over economy. However, if one wanted a diesel automobile, there were very few, expensive choices.

In 1973, the Arab oil embargo forced a dramatic increase in gasoline prices. Fuel economy was suddenly an important factor in a car. Manufacturers looked at alternative sources and concluded that the diesel engine had the capability to produce high mileage ratings without necessitating expensive or exotic technology.

In 1977, the first lightweight, diesel economy car for the United States was made by Volkswagen. General Motors, in the following year, produced a V-8 diesel for its full-sized cars. Today, there are at least 10 manufacturers producing a variety of diesel engines for a variety of car models and small trucks. Current information indicates that the diesel engine option will continue to expand over the next 10 years.

APPLICATION

The diesel is a versatile engine. It is capable of producing from 1 to 50,000 HP. This wide range allows the diesel to be used to power ships, generators, construction equipment, farm equipment, large trucks, as well as passenger cars and small trucks. Not only is the diesel adaptable, but it is also capable of running more economically and dependably than other power sources. Hence, diesels completely dominate areas such as the large truck class and construction equipment.

The diesel engine is designed for the specific application, thus creating several differences among diesel engines. The passenger car or small truck uses a high-speed diesel. The high-speed diesel differs significantly from the large slow-speed diesels used in large trucks. The high-speed diesel:

- has a higher RPM limit (approximately 5000 RPM vs. 2100 RPM)
- does not produce a high sustained torque curve
- uses a different combustion chamber
- uses different starting aids
- frequently uses different fuel systems

These differences are the result of the application, since the high-speed diesel is designed to give the same per-

Figure 1-5 Development of the diesel engine, in part, was advanced by the Cummins Engine Company, Inc., by racing diesel powered cars. Such cars qualified and ran in the 1931, 1934, 1950, and 1952 Indianapolis 500 mile races.

In 1931, a Cummins diesel-powered car completed the Indianapolis 500 without a pitstop.

In 1952, Cummins entered the Indianapolis 500 with an aerodynamic, low silhouette, turbocharged, diesel-powered race car. The engine produced 430 horsepower at 4,500 RPM (the stock engine at that time produced 125 horsepower at 2,500 RPM). It won the pole position with a new four-lap record of 138.010 mph. However, during the race, tire rubber dust from other cars gathered in the turbocharger, choking off the air supply. Still running, but smoking, the car was pulled from the race in the 72nd lap, and was never raced again.

formance and feel as a gasoline engine. It is not designed to haul heavy loads over an extended period of time. Other differences will become clear later.

THE DIESEL ENGINE AND THE GASOLINE ENGINE

The small high-speed diesel is often an option for new car and small truck buyers. The high-speed diesel engine has both advantages and disadvantages.

Advantages of the Diesel Engine

The basic advantages of the diesel engine are low fuel consumption (greater thermal efficiency), less fire hazard, and lower emission levels.

Low fuel consumption. The three primary factors for the diesel engine's low fuel consumption are air–fuel ratio, compression ratio, and low pumping losses.

The *air–fuel ratio* is the required amount of air to fuel needed to produce combustion. In the gasoline engine the air–fuel ratio is usually from 13:1 at idle to 17:1 under light cruise operation. The diesel ranges from 100:1 at idle to 20:1 under acceleration, figure 1-6. Only the amount of fuel needed is injected into the cylinder.

The *compression ratio* is the comparison of cylinder volume when the piston is at bottom dead center (BDC) to cylinder volume when the piston is at top dead center (TDC), figure 1-7. The formula is this:

$$\frac{\text{volume at BDC} + \text{volume at TDC}}{\text{volume at TDC}} = \text{compression ratio}$$

Today's gasoline engines have approximately an 8:1 compression ratio. The diesel ranges from 17:1 to 23:1,

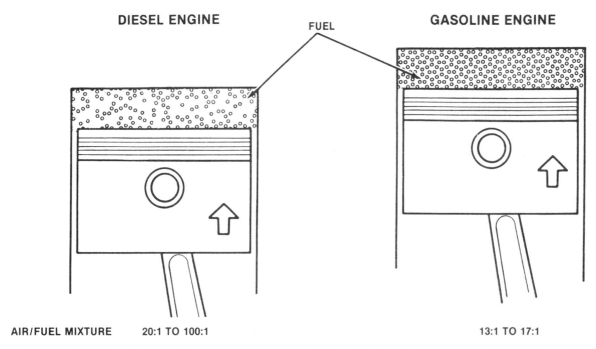

DIESEL ENGINE　　　　FUEL　　　　**GASOLINE ENGINE**

AIR/FUEL MIXTURE　　　20:1 TO 100:1　　　　　　　　13:1 TO 17:1

Figure I-6 The diesel engine can operate on a wide air/fuel ratio. *(Courtesy of Ford Motor Company)*

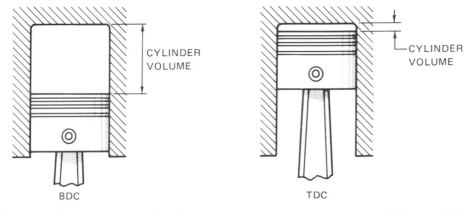

CYLINDER VOLUME

CYLINDER VOLUME

BDC　　　　　　　TDC

Figure I-7 A comparison of cylinder volume when the piston is at BDC and TDC

GASOLINE ENGINE

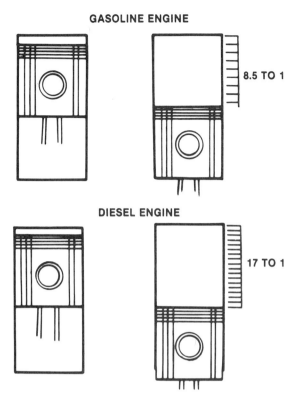

DIESEL ENGINE

Figure 1-8 The diesel engine operates on a much higher compression ratio. (*Courtesy of Ford Motor Company*)

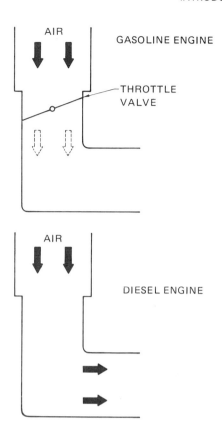

Figure 1-9 The diesel engine tries to pull in as much air on each intake stroke since there is no throttle valve.

figure 1-8. The compression ratios of small high-speed diesels used in passenger cars range from 21:1 to 23:1. The higher the compression ratio, the more efficient the engine will be because more energy is being extracted from the fuel.

Pumping loss is the amount of energy expended when pulling air into the cylinder and pushing the exhaust gases out. The gasoline engine has a throttle restriction (throttle valve in the carburetor or air valve with fuel injection), but the diesel engine does not, figure 1-9. The greater the restriction, the less efficient the engine is because it takes power to overcome the restriction. The diesel engine does not waste energy trying to pull air past the throttle valve; thus it has less pumping loss.

Compared to the gasoline engine at idle and deceleration, the diesel can be 40% to 50% more efficient. Also, very little or no fuel is injected into the cylinder under these conditions.

Very low pumping losses, very lean air–fuel ratios, and high compression ratios help make the diesel more *thermally efficient*. That is, of the total amount of energy available in the fuel, the diesel will convert more of the fuel's heat energy into usable power than the gasoline engine, figure 1-10. Another bonus is that diesel fuel contains more energy per gallon than gasoline, figure 1-

11. It takes less diesel fuel than gasoline to do the same amount of work.

Less fire hazard. Diesel fuel does not evaporate as readily as gasoline. The ease with which a liquid evaporates is called *volatility*. Gasoline is very volatile compared to diesel fuel and will ignite much more easily, figure 1-12. This is no reason, however, to handle diesel fuel any less carefully than gasoline!

Lower emission levels. The government has regulated three tailpipe emissions to date: hydrocarbons (HC), carbon monoxide (CO), and oxides of nitrogen (NOx). *Hydrocarbons* are present in fuel that has not been burned. *Carbon monoxide* is fuel that has been partially burned. Because the diesel can draw in a full amount of air and can run on very lean mixtures, it can burn the fuel more completely, thus changing the fuel and air more completely to carbon dioxide (CO_2) and water (H_2O), figure 1-13. Hence, there is no need to add an emission system to control HC and CO.

The diesel also produces less NOx. These *oxides of nitrogen* are formed when combustion temperature is above approximately 2500°F, at which oxygen and nitrogen combine. The diesel does not produce as much NOx as the gasoline engine because the peak cylinder temperature is not as high, slowing the chemical formation of NOx, figure 1-14. Some of the larger diesel engines do need a special device to lower NOx to meet government regulations.

Disadvantages of the Diesel Engine

The following disadvantges of the diesel engine can be attributed to the diesel's characteristics or to the subjective opinion of the owner.

Diesel engine construction costs are higher.
Because the diesel puts more stress on its components, the engine must be constructed from special materials. The quality of the material must be exact, and the parts must be fitted together with very little tolerance for error. This means assembly costs are higher. Special items must be added. For example, a heavier-duty starter system is needed. Extra sound-insulating material is needed, and the chassis must be strengthened to accommodate the extra weight. All these items increase the cost.

Different maintenance procedures.
Because of the diesel engine's design, service and maintenance procedures are different, and the technician must be more precise in making repairs. The fuel and starting systems require a new method of troubleshooting the diesel engine.

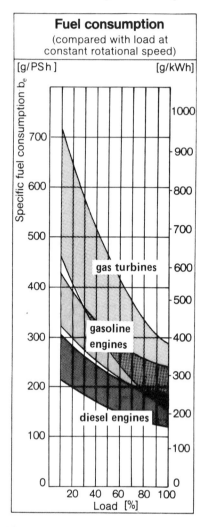

Figure 1-10 When compared to the gas turbine and gasoline engine, the diesel engine has the lowest fuel consumption. (*Illustration only Courtesy of Robert Bosch Corporation, Inc.*)

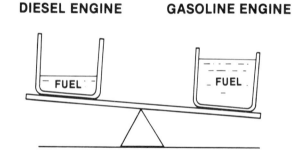

Figure 1-11 Since diesel fuel has more energy per gallon than gasoline, less diesel fuel will be consumed. (*Courtesy of Ford Motor Company*)

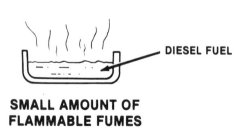

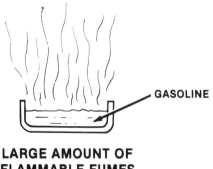

Figure 1-12 Diesel fuel is less volatile than gasoline. (*Courtesy of Ford Motor Company*)

REASON FOR LOW HC AND CO EMISSION

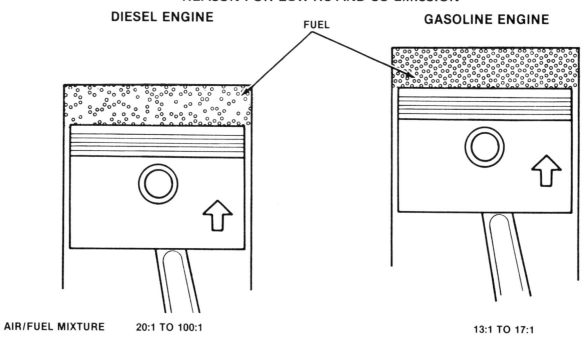

Figure 1-13 The high air/fuel ratio lowers HC and CO emissions. *(Courtesy of Ford Motor Company)*

REASON FOR LOW NO_x EMISSION

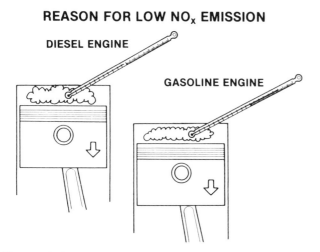

Figure 1-14 Peak combustion temperature in a diesel is not maintained long, lowering NO$_x$ emissions. *(Courtesy of Ford Motor Company)*

The following chapters will discuss these special service and maintenance procedures.

Cold weather starting. Diesels use the heat of compression to ignite the fuel. The colder the air temperature, the harder it is to build up enough heat for ignition. To aid cold weather starting, manufacturers have added special starting aid packages. These packages may include an engine block heater, special fuel heaters, a high-power starter system, and a glow plug system. These features are designed to add heat to the engine or provide extra starting power. For passenger car diesels, serious engine damage could result from use of ether or other starting aids. See Chapter 10 for further details.

Some tractor and large truck diesel engines may use ether or have an air preheater. The ether works because the compression ratio in the large diesels is generally lower than 19:1.

Engine noise. Diesel engines produce a knock, particularly at idle, that is very noticeable. This is due to the high compression and the nature of the combustion process. This knock is particularly noticeable in cold weather because of uneven combustion. Engineers try to limit this noise as much as possible by adding special sound barriers to the engine and chassis to prevent the noise from reaching the passenger compartment. For illustration, look at the underside of a diesel-powered car's hood. The thick insulating blanket is designed to dampen the noise.

As a technician you must listen to the different sounds the diesel makes and train your ear to pick out the abnormal noises.

Exhaust smoke and odor. Anyone who has followed a diesel, particularly one with a malfunction, knows that the diesel produces smoke and an odor all its own. The black smoke most often visible is soot (particulates). This is a result of the fuel not mixing and burning properly, or of insufficient air to complete the combustion process. The diesel emits some particulates even when operating normally. The government has stated that by 1985, automobile diesel engines must meet a particulate standard. Engineers are working on ways to solve this problem.

White smoke is formed when the engine is operated under low temperatures, light load, or excessive ignition delay. The white smoke is air mixed with diesel fuel that has not been burned. This smoke is also normal when first starting a diesel, but should clear up as the combustion chamber heats up.

A blue smoke is caused by excessive oil comsumption.

The odor, like the smoke, is a characteristic of the diesel engine and is due to engine design and the combustion process. Some odor, of course, is normal.

Low horsepower-to-weight ratio. The components of the diesel engine must be made durable enough to withstand the tremendous stress of the diesel combustion process. The diesel engine weighs more than the gasoline engine. Since the diesel engine weighs more, and sound-deadening insulation is needed, the chassis must be stronger and heavier to accommodate the extra weight. The top RPM of the diesel is limited because of the nature of the combustion process. This limits the amount of horsepower it can produce. When compared to the gasoline engine, the diesel engine has a low power-to-weight ratio. The horsepower-to-weight ratio can be increased by using light materials and by adding turbochargers, but these also increase cost.

SUMMARY

The diesel engine, dating back to the nineteenth century, is used in a wide variety of applications where durability and economy are paramount.

The small, high-speed diesel engine used in passenger cars and small trucks is different in design and purpose from the large slower-speed diesels. These differences require different service and repair procedures. The diesel has some advantages and disadvantages when compared to the gasoline engine. Whether or not the advantages outweigh the disadvantages depends on cost, oil market conditions, and government regulations.

CHAPTER 1 QUESTIONS

1. What is meant by the term *compression-ignition?*
2. List five differences between the high- and low-speed diesel.
3. List three advantages of the diesel engine.
4. Why does the diesel engine provide superior fuel economy?
5. Why does the diesel engine produce fewer emissions?
6. List the disadvantages of the diesel engine.

chapter 2
Fundamentals of Diesel Engine Operation

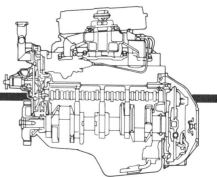

Objectives

In this chapter you will learn:
- The basic components of the diesel engine
- The four-stroke diesel cycle
- Valve timing
- The process and effects of compression
- The combustion process
- Load and speed control

BASIC ENGINE COMPONENTS

The diesel engine is an internal combustion, compression-ignition, heat engine. It is capable of converting the fuel's heat energy to *kinetic* (motion) *energy.*

The major components needed to perform the combustion process are the engine block, cylinder head, piston assembly, crankshaft, connecting rod, camshaft, valve train assembly, and flywheel, figure 2-1.

The engine block is the main structural member housing other components or having components fastened to it. The block provides the basic cylinder shape. At the top of the cylinder is the cylinder head, which is bolted tightly to the engine block. The cylinder head contains the intake

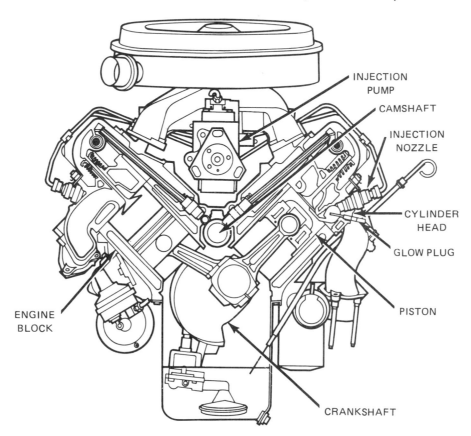

Figure 2-1 Major components in the combustion process *(Courtesy of Chevrolet Division)*

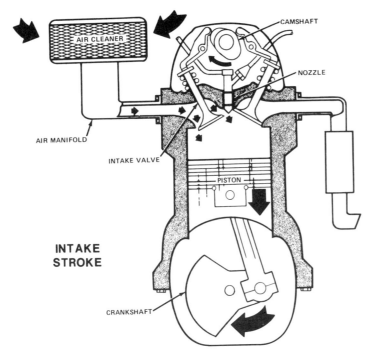

CAMSHAFT
NOZZLE
AIR CLEANER
AIR MANIFOLD
INTAKE VALVE
PISTON
INTAKE STROKE
CRANKSHAFT

Figure 2-2 Air is drawn in through the open intake valve as the piston travels downward. (*Courtesy of General Motors Product Service Training*)

and exhaust valves, a nozzle, and (on some engines) the camshaft.

The piston is the only part in the combustion area designed to move when combustion occurs. As it reciprocates (moves up and down) in the cylinder bore, it must maintain a tight seal and yet move with minimal frictional loss.

The piston is connected to a connecting rod, which is fastened to the engine crankshaft. The crankshaft converts the up-and-down movement of the piston into a circular (rotary) motion. Bolted to the end of the crankshaft is the flywheel. This is a heavy metal disc that stores kinetic energy. The flywheel is necessary to keep the engine turning when power is not being produced.

The valve train assembly opens and closes the intake and exhaust valves at the proper time, allowing fresh air to enter the cylinder and spent gases to exit.

These are the major parts in the four-stroke diesel cycle. Each stroke is the movement of the piston between BDC and TDC. The four strokes in order of occurrence are intake, compression, power, and exhaust. The following is how these four strokes generally work.

The Intake Stroke

Starting with the piston at TDC, the piston is pulled downward by the connecting rod through the movement of the crankshaft. As the piston is pulled downward, two other actions happen simultaneously. First, the camshaft opens the intake valve. Second, air is drawn past the intake valve by the downward movement of the piston. The diesel engine draws in as much air as possible since there is no throttle restriction. When the piston reaches BDC, the intake stroke is completed, figure 2-2.

The Compression Stroke

The piston travels upward as the crankshaft continues to rotate. The camshaft has closed the intake valve, so the cylinder is sealed and the air trapped. As the piston moves upward, the air becomes compressed and temperature increases to approximately 1000°F. Just before TDC, the nozzle begins to spray fuel into the hot compressed air, figure 2-3.

The Power Stroke

The power stroke is sometimes called the expansion stroke. Fuel, injected into the hot, highly compressed air, mixes with the air and burns rapidly. The combustion of the air–fuel mixture liberates the fuel's heat energy, which causes a tremendous rise in heat and pressure in the

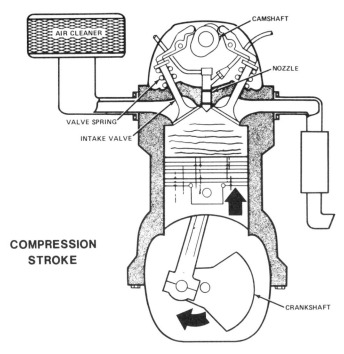

Figure 2-3 The trapped air is compressed as the piston travels upward. Just before TDC, fuel is injected into the hot air and combustion begins. (*Courtesy of General Motors Product Service Training*)

combustion area. It is this heat and pressure that provides the power to move the engine components, and the vehicle. The high pressure forces the piston, the only part of the cylinder area that can move downward. As the piston moves downward, cylinder volume increases. This causes the pressure to decrease, and the burning gases to expand. As the burning gases expand, their temperature decreases, figure 2-4.

The Exhaust Stroke

The cylinder must be purged of the spent gases. The piston moves from BDC upward accomplishing this task. At the same time, the camshaft has opened the exhaust valve. This allows the piston to push the spent gases into the exhaust system. After the exhaust gases are pushed out, the exhaust valve closes, figure 2-5. After the piston reaches TDC, the diesel cycle is repeated, starting with the intake stroke. Note that the flywheel has provided the necessary momentum (kinetic energy) to keep the engine turning during the intake, compression, and exhaust strokes.

VALVE TIMING

As previously stated, the diesel engine draws in a full charge of air on every intake stroke. How much of the cylinder fills with air is measured by a percentage called the *volumetric efficiency*. For example, if only 80% of

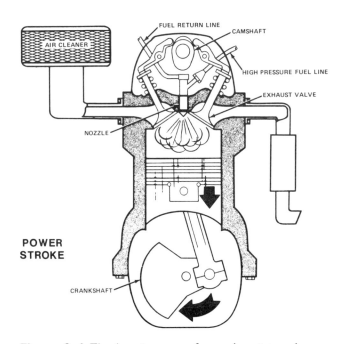

Figure 2-4 The burning gases force the piston downward. (*Courtesy of General Motors Product Service Training*)

the available cylinder volume is filled with air at standard pressure, its volumetric efficiency is 80%, figure 2-6. Volumetric efficiency is influenced by air pressure (density), air intake and exhaust restrictions, velocity of the air and exhaust gases, the camshaft, and engine speed. The camshaft is designed to provide good power and economy. It does this by opening and closing the intake and exhaust valves at the proper time. This is called *valve timing*. Valve timing is determined by the shape of the lobes on the camshaft.

Before discussing valve timing further, two other factors must be understood. First, as the piston travels through its stroke, its speed changes. Beginning at BDC, it has zero velocity, increases to a maximum velocity, and then decreases to zero velocity at TDC. There the piston stops

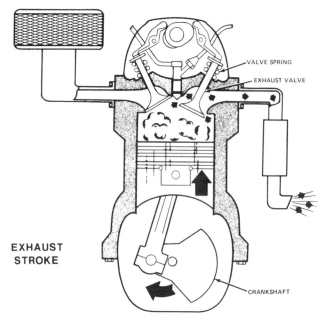

EXHAUST STROKE

Figure 2-5 The piston travels upward forcing the spent gases out the open exhaust valve. (*Courtesy of General Motors Product Service Training*)

and changes direction, increasing to maximum velocity and decreasing to zero velocity BDC. When the piston is near either TDC or BDC, its speed and movement are minimal, figure 2-7.

The second factor is that air and exhaust gases do not move instantaneously. It takes time to get these gases moving. Once these gases are under motion, they continue to move by their own momentum and eventually will stop if no other force acts upon them, figure 2-8.

Valve timing must make use of the varying piston speeds and the different speeds of the intake air and exhaust gases. Therefore, the four strokes of the diesel cycle, when measured in crankshaft degrees, are going to be of unequal length, figure 2-9.

Intake Valve Timing

The intake valve starts to open before TDC and closes after BDC. By opening the intake valve before TDC, fresh air can be drawn in early and the valve will be fully opened when the piston travels downward. The air is initially pulled in by the departing exhaust gases. On the exhaust stroke, the exhaust gases are being pushed out by the piston, but as the piston approaches TDC, its speed and movement decrease. However, most of the exhaust gases in the exhaust manifold depart from the cylinder with great speed. This helps create a suction or low-pressure area behind the exhaust gases. Therefore, when the intake valve opens just before TDC, fresh air is drawn in by the suction created by departing exhaust gases. The incoming fresh air helps scavenge the cylinder of any

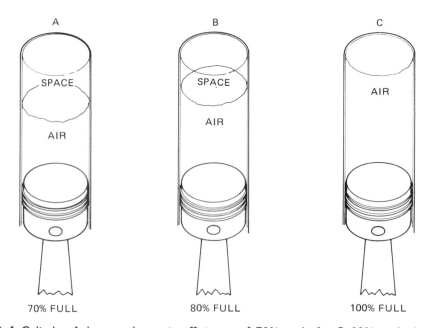

Figure 2-6 Cylinder A has a volumetric efficiency of 70%, cylinder B 80%, cylinder C 100%.

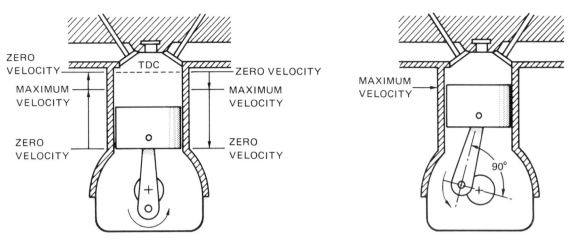

Figure 2-7 Piston speed changes as the angle between the centerline of the connecting rod and the center of the crankshaft changes. Maximum piston velocity occurs when this angle reaches 90°.

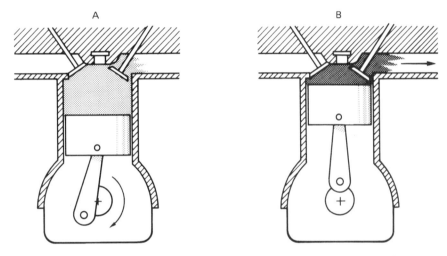

Figure 2-8 Exhaust gases resist being moved in cylinder A, but once under motion, they stay in motion though the piston has reached TDC as shown in cylinder B.

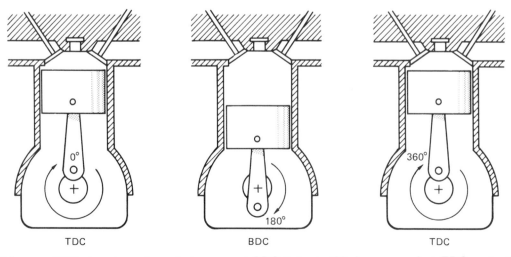

Figure 2-9 When at TDC the piston is at 0 degrees, at BDC it is at 180 degrees, and at TDC again the piston is at 360 degrees, or 0.

remaining exhaust gases and gives the cylinder more time to draw in a fresh charge of air, figure 2-10.

When the piston is moving downward on the intake stroke, air is drawn in. However, when the piston reaches BDC, its speed and movement are dimished. The air, on the other hand, is still rushing into the cylinder because of the momentum imparted to it by the piston. Keeping the intake valve open just after BDC packs the incoming air into the cylinder, figure 2-11.

Exhaust Valve Timing

To purge the cylinder of exhaust gases, the exhaust valve is timed to open before BDC and to close after TDC on the exhaust stroke. The opening of the exhaust valve just before BDC on the power stroke releases the remaining low pressure and ensures full exhaust valve opening by the time the piston does reach BDC. The remaining low pressure helps remove the exhaust gases and avoids putting pressure on the piston when it begins to travel upward, figure 2-12. Thus, what was lost in power by opening the exhaust valve early is regained by helping purge the cylinder and preventing back pressure on the piston. Keeping the exhaust valve open after TDC helps scavenge the cylinder completely. The departing exhaust gases create a suction that helps pull in fresh air for the intake stroke.

The time period in which both valves are open is called *valve overlap*, figure 2-13.

THE COMPRESSION PROCESS

Study of the compression process is important in understanding how the diesel ignites the fuel and converts the fuel's heat energy to kinetic energy. Furthermore, the high pressures developed as a result of compression strongly influence the construction, troubleshooting, and repair of the diesel engine.

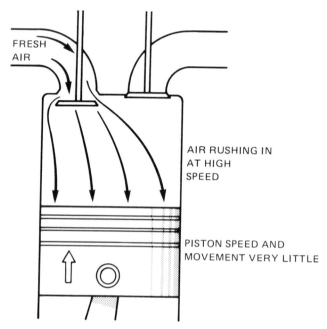

Figure 2-11 The intake valve is open even after the piston has reached BDC on the intake stroke.

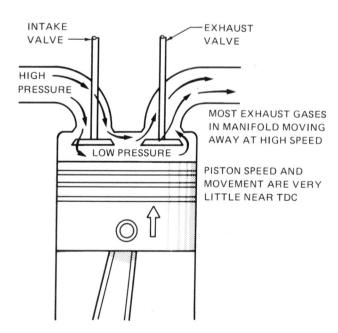

Figure 2-10 When the piston nears TDC on the exhaust stroke, the intake valve starts to open.

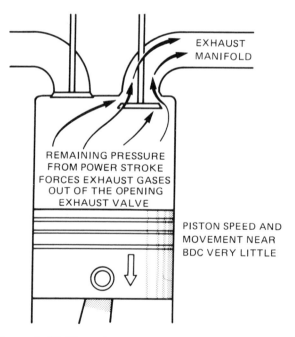

Figure 2-12 The exhaust valve opens before the piston reaches BDC on the power stroke.

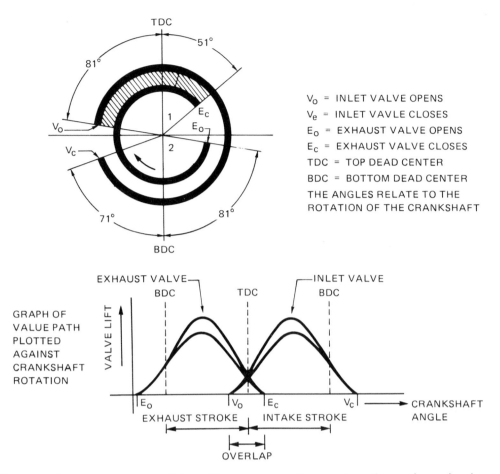

Vo = INLET VALVE OPENS
Ve = INLET VAVLE CLOSES
Eo = EXHAUST VALVE OPENS
Ec = EXHAUST VALVE CLOSES
TDC = TOP DEAD CENTER
BDC = BOTTOM DEAD CENTER
THE ANGLES RELATE TO THE
ROTATION OF THE CRANKSHAFT

Figure 2-13 Both graphs represent valve timing. The top graph shows when the intake and exhaust valves open and close as measured by crankshaft degrees. The bottom graph shows valve lift, and opening and closing in relation to piston position. If the intake and exhaust strokes are measured in crankshaft degrees, they would be of longer duration than the compression and power strokes. Keeping the valves open longer than one piston stroke improves volumetric efficiency. (*Courtesy of General Motors Product Service Training*)

The purposes of the compression stroke are (1) to provide enough heat to ignite the air–fuel mixture, and (2) to provide a means of extracting the fuel's heat energy.

Air is compressed when the cylinder volume is reduced, which occurs when the piston travels upward on the compression stroke. When the volume decreases, the compressed air's temperature and pressure rise. This increase is caused by the reduction in space between the air molecules. The molecules, at that time, do not have very far to go before colliding with each other. The greater the activity of the molecules, the greater the temperature and pressure. The greater the reduction in space between the molecules, the higher the temperature and pressure, figure 2-14.

The compression ratio (CR) is an important factor. The greater the CR, the higher the compressed air's temperature and pressure. To achieve the higher temperature

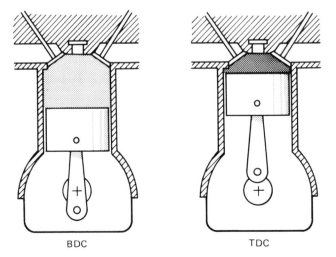

Figure 2-14 Air molecules are far apart at BDC, dense at TDC.

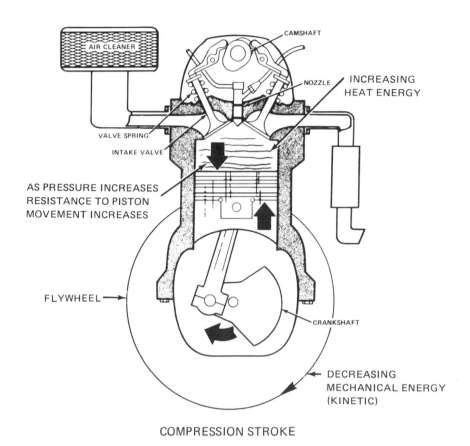

COMPRESSION STROKE

Figure 2-15 As the piston compresses the air, the air temperature rises or increases in heat energy.

and pressure, the piston must apply greater force on the air. The piston gets its force or energy from the flywheel. The piston is driven upward by the kinetic energy (momentum) stored in the flywheel. As air temperature and pressure increase, the amount of kinetic energy needed increases. Energy can neither be created nor destroyed, only the form energy takes may change. In this case, the kinetic energy stored in the flywheel is transmitted to the piston. As the piston works on the air, the kinetic

energy is changed to heat energy, figure 2-15. The greater the amount of kinetic energy forced on the air, the higher the heat energy.

The speed of the piston is another important factor. The faster the air is compressed, the more heat is retained by the air and less heat loss to the combustion chamber walls occurs, figure 2-16. If the piston travels far enough and fast enough in the cylinder bore, sufficient heat will be produced to start combustion.

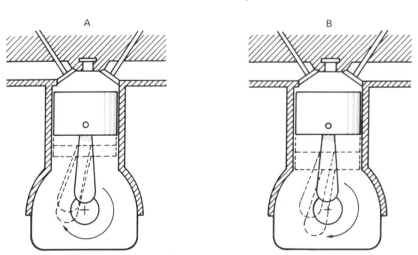

Figure 2-16 Cylinder B compresses the air faster retaining more heat than cylinder A.

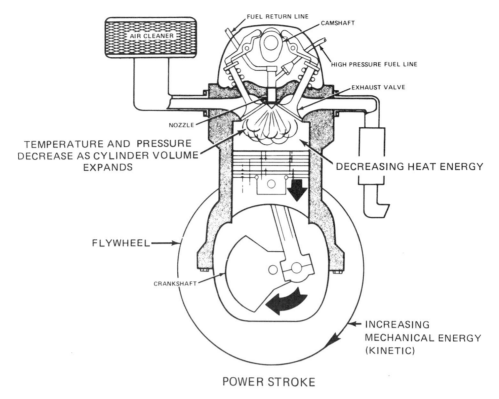

Figure 2-17 During the power stroke, the fuel's heat energy is liberated through combustion.

Compression also provides a means of extracting the fuel's heat energy. When the air–fuel mixture burns, the combustion chamber's temperature and pressure increase dramatically. This is because the fuel's heat energy is being released in a compressed state. The heat energy being released now works on the piston forcing the piston downward. The piston's downward movement is the result of heat energy converted to kinetic energy, figure 2-17. This is just the opposite of what happened during the compression stroke. (Remember, energy can neither be created nor destroyed, only the form may change.)

The Advantages of a High Compression Ratio

There are four main reasons why a high compression ratio yields a high thermal efficiency.

1. The higher the CR, the higher the *expansion ratio.* The expansion ratio, like the CR, is the comparison of cylinder volume from the time the piston is at BDC to when the piston is at TDC. But unlike the CR, the expansion ratio is the amount the burning gases expand when the piston is traveling downward during the power stroke. When combustion initially occurs, the combustion chamber volume is very small. As the burning gases force the piston downward, the combustion chamber volume expands. During this expansion, the burning gases cool off because the fuel's heat energy is converted to kinetic

energy. Increasing the expansion ratio increases the amount of heat energy that is to be converted into kinetic energy because of greater expansion, figure 2-18. If more heat energy is changed to kinetic energy, less heat is ejected into the exhaust system, and less heat energy is wasted. This is the reason the exhaust system on a diesel does not get as hot as a gasoline engine's exhaust system.

2. A high CR helps mix the air with the fuel. When the air is compressed, the air molecules are very active. This activity mixes more of the air molecules with the fuel molecules, causing more fuel to burn and release more heat energy, figure 2-19.

3. A high CR creates a small combustion space at TDC. A small combustion space does not allow as much heat to escape through the combustion chamber walls. More heat is retained during the compression and power strokes. This makes igniting the fuel easier as well as improving thermal efficiency, figure 2-20.

4. The higher the CR, the smaller the volume when the piston is at TDC. During the exhaust stroke, the piston pushes more of the spent gases out of the cylinder. This allows more fresh air into the cylinder, which provides better combustion, figure 2-21.

The engine that can make use of high compression ratios has a better thermal efficiency. Because the diesel engine draws in only air, no ignition is possible until fuel is added. Therefore, the diesel engine compresses the air

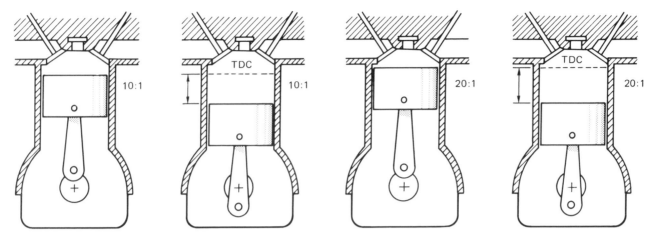

Figure 2-18 The cylinder with the 20:1 CR allows for greater expansion of the burning gases, converting more heat to power.

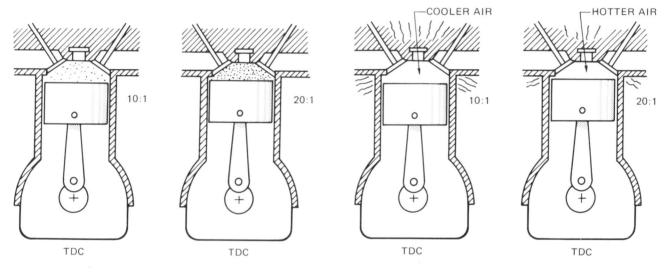

Figure 2-19 There is better distribution of fuel into the compressed air in the cylinder with the higher CR.

Figure 2-20 More heat is lost at cylinder A because of the greater combustion surface area at TDC.

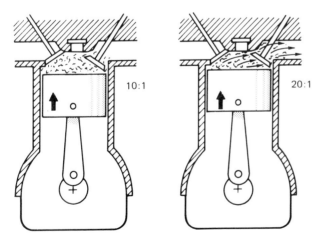

Figure 2-21 More of the exhaust gases will be pushed in the cylinder with the higher CR.

as much as possible to obtain maximum efficiency. The gasoline engine is limited because the fuel has been previously mixed with the air and this mixture ignites if compressed too much, causing harmful detonation, figure 2-22.

The Limitations of High Compression Ratios

There are limits on how much the air can be compressed in the diesel engine.

1. As the CR is increased, the gain in power becomes less and less. Finally, a point is reached when any increase in CR does not result in any gain of power. This is due to the reduction in combustion volume in relation to piston movement being so small that any expansion in relation

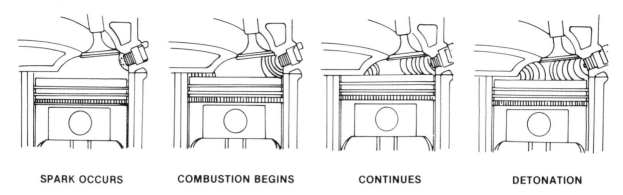

SPARK OCCURS   COMBUSTION BEGINS   CONTINUES   DETONATION

Figure 2-22 In a gasoline engine, if the compression ratio is too high or the octane of the gasoline too low, ignition occurs at a point other than at the spark plug. *(Courtesy of Ford Motor Company)*

to piston movement is negligible. Therefore, the resultant power gain is negligible, figure 2-23.

2. A high CR can reduce *mechanical efficiency* because of friction losses. When the CR is increased, the stress on the engine components becomes greater, and tighter seals are necessary. The bearing surfaces, pistons, crankshafts, connecting rods, and so on must all be strengthened to withstand the greater loads. These heavier parts decrease power output because of their weight and greater surface contact areas. These larger surfaces areas increase the amount of friction between parts. More power must be produced just to keep these parts moving, decreasing engine power output.

3. A high CR requires a heavier-duty starting system to overcome the high compression pressures and to rotate

the heavier parts. The batteries must have enough power to supply the starting motor. The starter motor must be of sufficient size to rotate the engine fast enough. This adds extra weight to the vehicle.

Most small, high-speed diesel engines have a CR of approximately 21:1 or higher. This is very high, even for a diesel engine. The reason is that the type of combustion chamber used (called indirect injection) has a large combustion surface wall area that can conduct more heat away. To overcome this, manufacturers increased the CR. However, the CR is now so high that it leads to increased friction losses, particularly at high engine speeds. At the upper end of the high-speed-car diesel's RPM range, the fuel consumption almost equals that of the comparable gasoline engine.

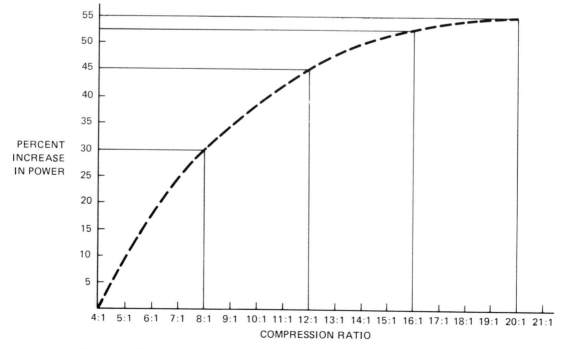

Figure 2-23 As the CR increases, the rate of power increase diminishes.

THE COMBUSTION PROCESS

Combustion takes place in the diesel when there is enough heat to ignite the correct proportions of air and fuel. Whenever the air and fuel are proportional, auto-ignition occurs. This means that combustion can start at more than one point, and that the combustion rate is determined by the rate at which the fast air and fuel mix. This process is called the *diffusion flame process.*

The combustion process in the diesel engine is divided into three periods. The first period is called the *delay time* or *ignition lag* and begins near the end of the compression stroke. This occurs when the fuel is first injected into the cylinder until the fuel begins to burn, figure 2-24. Time is needed to allow the fuel to vaporize and mix with the hot air. This period lasts about .001 second.

The second period is called the *rapid combustion period.* When the air–fuel mixture is proportional and the temperature high enough, auto-ignition occurs. The cylinder pressure rises suddenly when the first fuel is burned. This sudden rise in pressure is the familiar diesel knock you hear.

The third period is called the *controlled combustion period.* During this period, the nozzle is still putting fuel into the combustion chamber, but this rich core of fuel does not burn yet because it has not united with air. The already burning gases cause a tremendous turbulence, mixing the remaining air with the fuel. The cylinder pressure rate increase is slower than that of the previous period. It is during this period that the greatest release of energy takes place. As the fuel and air are consumed, cylinder pressure and temperature decrease as cylinder volume increases.

The rate of combustion in the diesel is slower than in the gasoline engine. It must be slower to allow enough time for the fuel to mix with the air, and burn in a controlled manner. As engine RPM increase, the amount of time available for combustion decreases. Because of the slower rate of combustion, the diesel engine cannot achieve the same high RPM as the gasoline engine.

The engine designer pays close attention to these three periods of diesel engine combustion, because they affect engine performance. A long ignition delay time (approximately .002 second) will allow more fuel to enter the combustion chamber. When this fuel does burn (the rapid combustion stage), the sudden rise in pressure is too much, causing the knock to be much louder. Note that knock in the diesel engine occurs at the beginning of the combustion process. Any factor that increases the delay time increases the knock. The tendancy to knock increases:

● When the air and engine temperature is low. The engine has a harder time building up enough heat to ignite the fuel at the proper time. If ignition occurs later than specified, more fuel will have entered the cylinder, causing the rise in pressure to be greater.

● When poor mixing of the air with the fuel occurs. This can be caused by the nozzle not spraying the fuel into the cylinder in the proper pattern.

● With the use of direct injection combustion chambers (this is discussed in chapter 3). This is one reason why indirect combustion chambers are used in passenger car diesels.

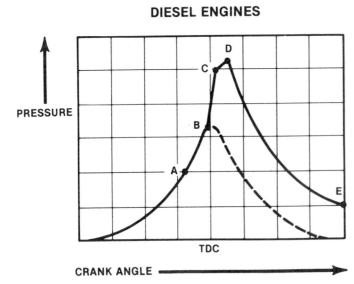

DIESEL ENGINES

DIESEL ENGINE:
A-D: Fuel injection time
A-B: Retarded time of ignition (formation of igniting conditions)
B-C: Flame transmission time (increase of pressure)
C-D: Direct combustion time (late burning)
D-E: Later combustion time (combustion of unburned gas)

Figure 2-24 The graph indicates the change in cylinder pressure during the different phases of the combustion. (*Courtesy of Ford Motor Company*)

● When a low cetane-rated fuel is incorrectly used. Diesel fuel is rated by the cetane method. The lower the cetane number, the longer the ignition delay period.

These are some of the factors affecting diesel knock and the combustion process of the diesel engine.

POWER AND SPEED CONTROL

Power and speed output in the diesel engine are controlled by the amount of fuel injected into the cylinder. The fuel system of the diesel engine injects enough fuel to meet the demand. At idle, the amount of fuel injected is small. When the demand is great, such as climbing a steep hill, the amount of fuel injected is greater and the combustion period longer. At maximum power, too much fuel in proportion to the amount of air may be injected into the cylinder. The remaining unburned fuel is heated, changing into carbon. This is called the *smoke limit.* It is reached whenever maximum power is needed. To shut off a diesel, cut off either the air supply or fuel supply. Shutting down the engine by shutting off the fuel supply is the standard method.

SUMMARY

The diesel engine uses four strokes to convert heat energy to kinetic energy. The intake stroke pulls in a fresh air charge. The compression stroke compresses the air to a high temperature and pressure. Fuel is injected into the hot air, and, after a beief period, the fuel and air ignite. The power stroke begins with the burning gases forcing the piston downward. As the cylinder volume expands, the burning gases cool, because the fuel's heat energy has been converted to mechanical energy. The exhaust stroke forces the spent gases out, and the diesel cycle is repeated, beginning with the intake stroke.

How well the engine pulls in air and expels the spent gases is strongly affected by valve timing. Opening the intake valve before TDC allows the exhaust gases to draw in fresh air. Closing the intake valve after BDC allows the fresh air to continue moving into the cylinder by virtue of its momentum. Opening the exhaust valve before BDC releases the remaining low pressure, which helps push out the spent gases and ensures that the valve is fully open when the piston travels upward. Closing the exhaust valve after TDC allows the remaining gases to escape and to help pull in a fresh air charge.

Compression serves to heat the air for fuel ignition and provides a means of extracting the fuel's heat energy. As the piston moves upward on the compression stroke, kinetic energy stored in the flywheel is converted to heat energy. The air is heated to a temperature that ignites the fuel. As the fuel burns, its heat energy is converted to kinetic energy.

Compression ratio directly influences thermal efficiency; the higher the CR, the higher the expansion ratio. The more a gas expands, the more heat can be extracted from it. A high CR improves fuel and air mixing, helps the air retain more heat, and purges the cylinder of exhaust gases. As the CR increases, the gain in power diminishes. Also, a high CR requires stronger materials, closer tolerances, and a heavy-duty starter system.

The combustion process is divided into three periods: the delay time, the rapid combustion, and the controlled combustion period. The delay time allows the fuel to mix with the air. The rapid combustion period occurs when the fuel first ignites, causing a steep rise in cylinder pressure. The controlled combustion period occurs when the remaining air and fuel mix, creating a controlled release of energy. The diesel combustion process is slower than the gasoline combustion process, thus limiting the RPM range of the diesel.

Power and speed control in the diesel depend on how much fuel is injected into the cylinder.

CHAPTER 2 QUESTIONS

1. Name the four strokes in their order of occurrence.
2. Name four factors affecting volumetric efficiency.
3. Why does piston speed change?
4. Where is piston speed greatest? Where is it slowest?
5. For what purpose is the intake valve opened before TDC and closed after BDC?
6. For what purpose is the exhaust valve opened before BDC and closed after TDC?
7. Why is compression important?
8. Why does increasing the compression ratio improve thermal efficiency?
9. Name and explain the three stages in the combustion process.
10. How are power and speed controlled in a diesel engine?

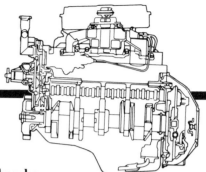

chapter 3
Combustion Chamber Designs and Their Influence on Engine Operation

Objectives

In this chapter you will learn:
- **The operation and characteristics of indirect combustion chambers**
- **The operation and characteristics of direct combustion chambers**
- **The characteristics of the high-speed diesel**

Although each engine component is important, the shape of the combustion chamber greatly determines the characteristics of the diesel engine. Regardless of the shape of the combustion chamber, this area must promote good combustion, limit emissions, limit noise, provide good fuel economy and, smooth operation.

The diesel engine combustion process is strongly influenced by air turbulence created by the shape of the combustion chamber area. The fuel injector or injection nozzle sprays the fuel into the combustion chamber in a pattern that takes advantage of the air turbulence, creating better mixing of the fuel with the air. Each combustion chamber shape creates its own unique air turbulence pattern that is right for some applications and wrong for others. It is important to know the different designs, although at present, some are not found on passenger car diesel engines.

Combustion chambers can be divided into two basic groups, based on the point where fuel is injected. The two categories are *direct injection* (DI) and *indirect injection* (IDI). DI is used in large, slow-speed diesels such as tractors and large trucks. Although DI is not used in small, high-speed diesels, it is important to know why this is so, and the advantages of this method.

DIRECT INJECTION

The DI method injects fuel directly into the combustion area above the piston. This method is often called the open combustion chamber, because the combustion chamber has direct access to the intake and exhaust valves, figure 3-1.

Getting the air to flow in the proper pattern within the combustion chamber is critical. The first step toward this goal involves how the air enters the cylinder at the intake valve. The shape of the intake passage directs the air in a manner that causes the air to swirl, figure 3-2. The second step involves the shape of the piston. Piston shape creates further turbulence as the piston rises on the compression stroke. There are different shapes to promote different patterns, figure 3-3. Combustion causes even more turbulence. All this air turbulence has been created to ensure proper mixing of air with fuel for good combustion.

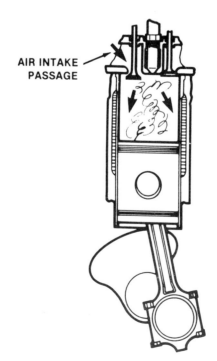

Figure 3-1 Direct injection combustion chamber (*Courtesy of Ford Motor Company*)

Characteristics of Direct Injection

The DI chamber has the highest fuel efficiency rating when compared to other designs. Its thermal efficiency is high primarily because there is little combustion surface

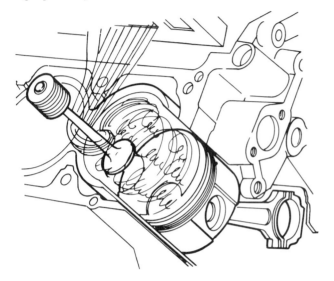

Figure 3-2 Intake air being directed and swirled as it enters the combustion chamber (*Courtesy of Ford Motor Company*)

wall area compared to combustion volume; therefore little heat loss occurs, figure 3-4. Lower compression ratios of approximately 15:1 to 18:1 can be used. (Remember from Chapter 2 that above 15:1 the power gain is minimal and frictional losses offset any gain.) Because of the low heat loss, a diesel engine with DI will start more easily and will not require extensive starting aids found in other types of combustion chambers. Finally, with DI the cylinder is easily purged of exhaust gases.

DI is not used in passenger car and small truck applications because it has a limited RPM range, up to approximately 2500 RPM.

The upper RPM limit is determined by the ability to take enough air into the cylinder and mix it properly with the fuel. Because the engine may not take in enough air at high RPM, the fuel will not mix properly, causing engine misfires that produce more hydrocarbons and particulates. Current U.S. government emission regulations are too stringent to permit the use of DI engines in passenger cars. When engine RPM increase, there is less time for the cylinder to take in air, and volumetric efficiency decreases. The engine designer overcomes this by using bigger valves, but with the use of DI the valves and nozzle are competing for the same space. Space is

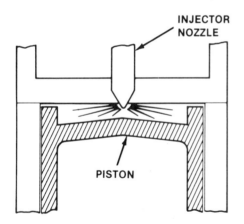

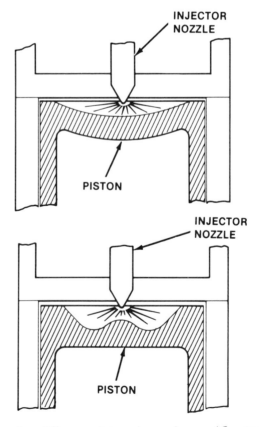

Figure 3-3 Three different types of open combustion chambers using different piston dome shapes. (*Courtesy of Ford Motor Company*)

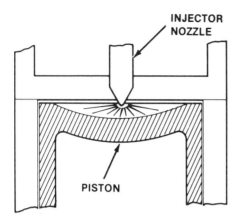

Figure 3-4 Low combustion chamber surface wall area to cylinder volume allows little heat loss. (*Courtesy of Ford Motor Company*)

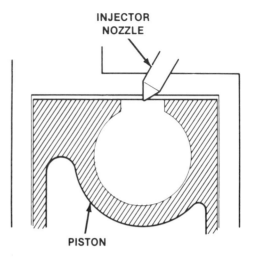

Figure 3-5 The M-system combustion chamber shows the most promise for a high-speed DI engine. (*Courtesy of Ford Motor Company*)

further limited by the relatively small cylinder bore diameter found in passenger car diesels. This again limits the amount of space that can be used by the valves.

Another characteristic of DI is that in rapid combustion phase cylinder pressure rises at an extreme rate. The sharp knocking noise produced would be considered objectionable by many car owners.

For these reasons, DI is not found in applications requiring a high engine RPM. But, because the fuel efficiency is so good, engineers are working to improve this design, figure 3-5. This will lead to an impressive mileage gain (about 10% to 15%) in fuel economy over today's diesel-powered cars.

INDIRECT INJECTION

IDI injects fuel into an antechamber that is connected to a main chamber by a narrow passage. This design is also called the divided chamber. There are no intake or exhaust valves in the antechamber. Air is pushed through the narrow passage on the compression stroke and becomes turbulent within the antechamber. After the nozzle sprays fuel into the antechamber, combustion begins in this area. However, there is not enough air in the antechamber to complete the burning of the fuel. The expanding burning gases force their way out into the main chamber and intensely mix with the remaining air. This gives a fast, complete burn of the remaining rich air–fuel mixture, figure 3-6.

There are variations of two basic IDI designs used in automobile and small truck applications: the precombustion chamber and the swirl chamber.

The Precombustion Chamber

The precombustion chamber is connected to the main chamber by a narrow passage. The precombustion chamber

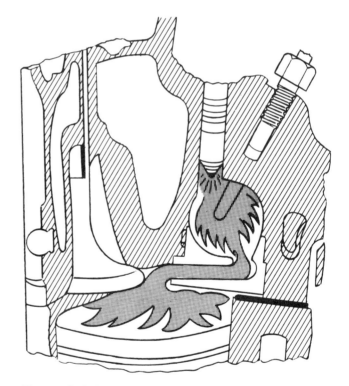

Figure 3-6 The expanding, burning air/fuel mixture rushes out into the main combustion area. (*Courtesy of General Motors Product Service Training*)

contains approximately 30% of the combustion volume when the piston is at TDC, figure 3-7. Air is forced through the narrow passage and becomes turbulent. Fuel is injected and the burning gases force their way through the narrow passge. This narrow passage speeds up the expanding gases more. The expanding gases mix with the air in the main chamber, rapidly completing the combustion

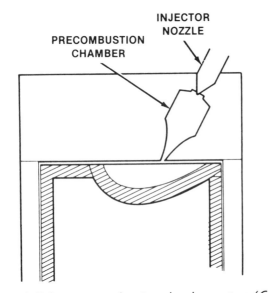

Figure 3-7 Basic precombustion chamber system (*Courtesy of Ford Motor Company*)

process, figure 3-8. There are variations on using this basic design.

The Swirl Chamber

This design is very similar to the precombustion chamber. However, the swirl chamber is spherical in shape and contains approximately 70% of the combustion volume, figure 3-9. Air is forced through the narrow passage into the swirl chamber. Here the air swirls, or becomes turbulent. Fuel is injected into the swirling air. Like the precombustion chamber, the burning mixture forces its way out the passage into the main chamber, where it completes the burning process.

It should be noted that the swirl chamber and precombustion chamber are very much alike, and manufacturers have used both names to describe the same chamber.

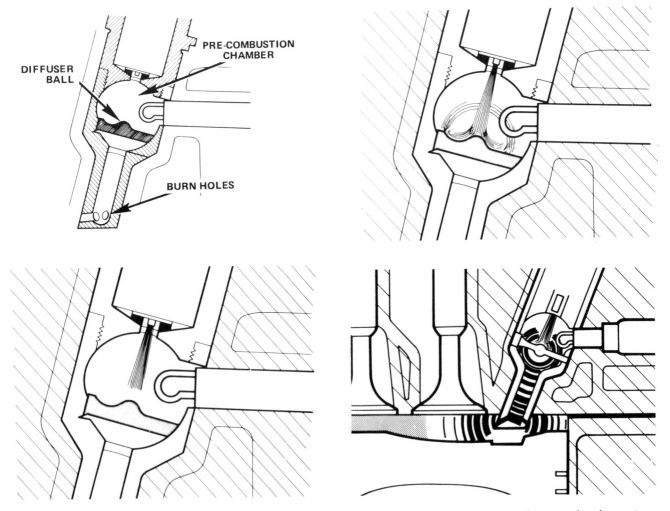

Figure 3-8 From top to bottom beginning with left column, fuel is injected into the precombustion chamber, mixes with the air, then ignites traveling out of the burn holes. (*Courtesy of Mercedes-Benz of North America Inc.*)

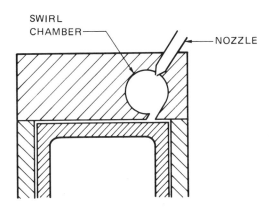

Figure 3-9 Basic swirl combustion chamber

Hybrid Chambers

Some designs incorporate features of the precombustion chamber and the swirl chamber. The shape of the combustion chamber strongly influences engine operation, noise, economy, and emissions. To meet these requirements, the engine designer uses certain features of each design. The result is a chamber that has characteristics of both the precombustion and swirl chamber, figure 3-10.

Characteristics of Indirect Injection

The characteristics of IDI have made it the logical choice for diesel engines in passenger cars. IDI provides wide RPM range, low emissions, and low noise.

Wide RPM range. IDI allows the diesel engine to operate up to 5000 RPM. There are two reasons for this. First, mixing the air with the fuel is not as difficult when using a precombustion chamber or a swirl chamber. The air becomes turbulent easily in these smaller, confined areas.

Second, use of a precombustion chamber or swirl chamber allows the use of larger intake and exhaust valves. With these chambers, the nozzle is not located directly above the piston. Therefore, the nozzle is not competing with the intake and exhaust valves for the same place. The larger intake and exhaust valves improve the volumetric efficiency of the engine, particularly at higher RPM when the engine has less time to breathe.

Low emissions. The fast, complete burn provided by IDI produces very little hydrocarbons, carbon monoxide, and soot. The injected fuel combines with the air and changes mostly to carbon dioxide and water as a result of the burning process. Oxides of nitrogen (NOx) are produced when a high peak temperature is maintained. IDI has a lower peak temperature than DI. The lower temperature does not allow the oxygen and nitrogen molecules to combine easily and form NOx.

Low noise. IDI does not produce the extreme rise in pressure when the fuel is first burned. Because pressure does not rise at an extreme rate, and combustion begins in a relatively small area, the noise produced is diminished when compared to a diesel with DI.

Disadvantages of Indirect Injection

The disadvantages of IDI are that it provides less fuel economy, and, requires an extremely high compression ratio.

Fuel economy is lower with IDI because of higher heat loss, higher friction losses, small pumping loss of gases in and out of the antechamber, and prolonged and retarded energy release.

The higher heat loss is the result of having a higher combustion surface wall area to combustion volume, figure

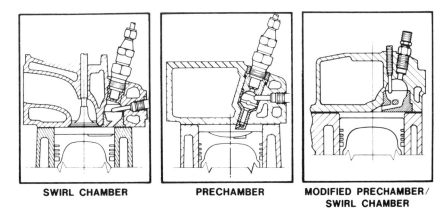

SWIRL CHAMBER   **PRECHAMBER**   **MODIFIED PRECHAMBER/ SWIRL CHAMBER**

Figure 3-10 A comparison of the swirl chamber, prechamber, and hybrid or modified prechamber/swirl chamber *(Courtesy of Ford Motor Company)*

3-11. This conducts more heat away from the combustion area, reducing thermal efficiency. Because heat losses are higher, a higher compression ratio is needed to increase the temperature, particularly for engine starting and light load operation. When starting the diesel a small amount of heat loss may prove the difference between a start and a no-start condition. Glow plugs are added to provide the additional heat when starting. Under light load operation, little heat is generated because little fuel is put into the cylinder. Having a high CR corrects these conditions, but the gain in thermal efficiency is very little. Furthermore, because a high CR is used, heavier parts and tighter seals are needed; therefore, friction losses are greater. This loss becomes greater as RPM increase, because it takes more power to keep these parts moving. Fuel economy decreases, particularly at high engine speeds.

The narrow passage restricts the flow of gases in and out of the antechamber. This restriction creates a small loss in pumping efficiency and scavenging ability.

The energy release is retarded with IDI, producing a slow rise in pressure and heat, converting less heat energy to mechanical energy. However, this slow rise is what gives IDI its low-noise and low-NOx characteristics.

THE CHARACTERISTICS OF THE AUTOMOBILE DIESEL ENGINE

The diesel engine is designed to share many of the same characteristics as a gasoline engine for use in automobiles. Typical characteristics of a gasoline engine are wide RPM range, ease of starting, smooth and quiet operation, and good acceleration.

Manufacturers wanted the diesel-powered vehicle to provide the same feel as the gasoline-powered vehicles. To meet these requirements, manufacturers chose IDI. A wide RPM range is essential for current transmissions. Because the RPM range is limited in DI engines, these

vehicles (such as large trucks) typically have 10- to 15- speed transmissions. Many automobile owners find this impractical. Even with IDI, the diesel engine still does not reach the same peak RPM as a gasoline engine. Hence, the diesel-powered car has less horsepower.

IDI provides smooth, quiet operation in comparison to the traditional diesel. It is still not as quiet as a gasoline engine, particularly at idle. Extra sound-deadening insulation is used in the hood and firewall to limit noise in the passenger compartment.

The gasoline engine accelerates faster than the diesel engine. However, one should keep in mind that most drivers do not require full throttle, so the acceleration of the diesel is adequate. To acquire higher performance levels, the diesel is easily adaptable to turbocharging. Some manufacturers use this adaptation.

The gasoline engine starts in a wide temperature range. The automobile diesel with IDI requires extensive starting aids to start easily. The diesel owner must also be sure that the fuel is blended for the proper temperature.

The big advantage of the diesel is the superior economy it offers. The fuel mileage, particularly in city driving conditions, is greater than for the comparable gasoline-powered model. Regarding maintenance, the diesel offers superior benefits. Generally, as long as the diesel receives clean fuel, air, and proper engine oil changes, it is cheaper to operate.

SUMMARY

The combustion chamber promotes good combustion, limits noise and emissions, provides good fuel economy and smooth operation. The combustion process is strongly influenced by how the air mixes with the fuel. There are several combustion chamber shapes and nozzle spray patterns used to accomplish this purpose.

DI places the nozzle directly above the piston. This method is used for applications where the engine speed is relatively slow. Above approximately 2500 RPM, misfires occur and emission levels increase. DI offers superior fuel economy, but its noise and emission levels are high, and it is hard to adapt if the cylinder bore diameter is small. For these reasons, DI is not used in car and small truck applications.

IDI is the preferred method for cars and small trucks. Because the combustion chamber is divided into smaller chambers, the air and fuel mix readily over a wide RPM range. Emission and noise levels are lower. However, extremely high compression ratios are needed to generate enough heat for engine starting and light load operation. Glow plugs are needed to provide extra heat when starting. Fuel economy is not as good as with DI because of higher

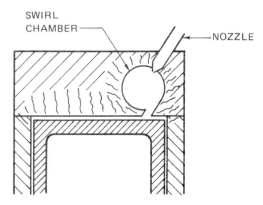

Figure 3-11 High combustion chamber surface wall area to cylinder volume has a greater heat loss.

friction losses, lower thermal efficiency, and a small pumping loss in and out of the main chamber.

The diesel engine is directly compared to the gasoline engine in terms of performance and economy because both engines are offered to the public in the same style of chassis. Manufacturers have designed their diesel-powered models to provide the same feel as their gasoline-powered counterparts. Still, the diesel-powered vehicles are slower in acceleration and have lower top speed. The slow acceleration is due to the increased weight of parts and the nature of the diesel combustion process. The top speed is lower because the maximum RPM output is limited to prevent engine misfire and damage. Regarding fuel economy, the diesel comes out on top, particularly at idle and light load operation. Maintenance economy can be greater as long as the engine receives clean fuel, air, and oil.

CHAPTER 3 QUESTIONS

1. Name five tasks the combustion chamber must perform.
2. Describe the operation of a DI combustion chamber.
3. Why is DI not used on diesel-powered passenger cars?
4. Describe the operation of IDI combustion chambers.
5. Why is IDI used on diesel-powered passenger cars?
6. What are the disadvantages of IDI?

chapter 4
Diesel Fuel

Objectives

In this chapter you will learn:
- Five differences between diesel fuel and gasoline
- The effects of cold weather on diesel fuel
- How cold weather's effects on diesel fuel are compensated
- The difference between cetane and octane ratings
- How ignition lag time is influenced by the cetane rating
- Why diesel fuel and gasoline should never be mixed
- Why a wound contaminated with diesel fuel should be treated immediately
- Five rules for storing diesel fuel

DIESEL FUEL CHARACTERISTICS

Diesel fuel, like gasoline, is made from petroleum. However, at the refinery, the petroleum is separated into three major components—gasoline, middle distillates, and all remaining substances.

Diesel fuel comes from the middle distillate group, which has properties and characteristics different from gasoline. Each of these characteristics will be discussed and contrasted with gasoline.

Heat Energy

Diesel fuel contains more heat energy than gasoline. The heat energy or value is commonly measured in *British thermal units* (BTU). One BTU is the amount of heat energy needed to raise the temperature of 1 pound of water 1 degree Fahrenheit, figure 4-1. (The metric equivalent of the BTU is the calorie. One calorie will raise 1 gram of water 1° Celsius.) The diesel engine converts the fuel's heat energy into power. If the fuel used has a high

heat energy content, more heat energy will be released. Hence, if two engines are identical, each having the same thermal efficiency, but are fed two different fuels, the engine that receives the fuel containing the higher BTU content would be more economical. It would produce the same power using less fuel.

Specific Gravity

The *specific gravity* of a liquid is a measurement of the liquid's weight compared to water. Water is assigned a value of 1. Diesel fuel is lighter than water but heavier than gasoline, and can change if it is mixed with other fuels. The specific gravity of diesel fuel is important to engine operation. The fuel must be heavy enough to achieve adequate penetration into the combustion chamber. If the specific gravity is too low, all the fuel immediately burns upon entering the combustion chamber. This puts all the force of combustion on one small area of the piston instead of equal force across the dome, figure 4-2. As a result, performance suffers, engine noise increases, and the piston could eventually be damaged.

Wax Appearance Point and Pour Point

Temperature affects diesel fuel more than it affects gasoline. This is because diesel fuels contain paraffin, a wax substance common among middle distillate fuels. As temperatures drop past a certain point, wax crystals begin to form in the fuel. The point where the wax crystals appear is the *wax appearance point* (WAP) or *cloud point*. WAP may change as a result of the origin of the crude oil and the quality of the fuel. The better the quality, the lower the WAP. As temperatures drop, the

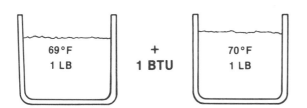

Figure 4-1 One British thermal unit of heat will raise the temperature of 1 pound (pint) of water 1 degree Fahrenheit. *(Courtesy of Ford Motor Company)*

INJECTION NOZZLE

Figure 4-2 Combustion is not spread evenly through the combustion chamber. (*Courtesy of Ford Motor Company*)

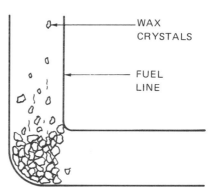

WAX CRYSTALS

FUEL LINE

Figure 4-3 Wax crystals collecting, plugging fuel flow as temperature drops

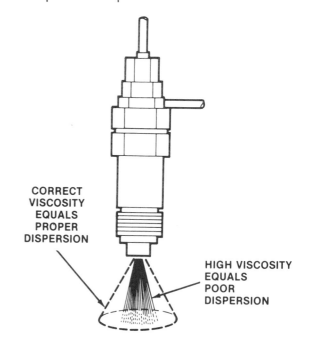

CORRECT VISCOSITY EQUALS PROPER DISPERSION

HIGH VISCOSITY EQUALS POOR DISPERSION

Figure 4-4 Fuel viscosity affects spray pattern. (*Courtesy of Ford Motor Company*)

wax crystals grow larger and restrict the flow of fuel through the filters and lines. Eventually, the fuel, which may still be liquid, stops flowing because the wax crystals plug a filter or line, figure 4-3. As the temperature continues to drop, the fuel reaches a point where it solidifies and no longer flows. This is called the *pour point*. In cold climates it is recommended that a low-temperature pour point fuel be used.

Viscosity

The viscosity of diesel fuel directly affects the spray pattern of the fuel into the combustion chamber and the fuel system components. Fuel with a high viscosity produces large droplets that are hard to burn. Fuel with a low viscosity sprays in a fine, easily burned mist, figure 4-4. If the viscosity is too low, it does not adequately lubricate and cool the injection pump and nozzles.

Volatility

Volatility is the ability of a liquid to change into a vapor. Gasoline is extremely volatile compared to diesel fuel. For instance, if diesel fuel and gasoline are exposed to the atmosphere at room temperature, the gasoline evaporates and the diesel fuel does not, figure 4-5.

Flash Point

Flash point is the lowest temperature at which the fuel burns when ignited by an external source. The flash point has little bearing on engine performance, but is important in fuel storage safety. (The temperature at which the flash point occurs is regulated.) If the flash point of diesel fuel were lower than specified, it would have the right combination of air and fumes that would ignite too easily, making the handling of it hazardous. Gasoline evaporates at a very low temperature, filling the tank with fumes that are potentially explosive.

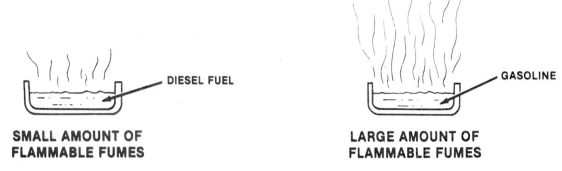

Figure 4-5 Gasoline is more volatile than diesel fuel, evaporating easily. *(Courtesy of Ford Motor Company)*

Cetane Rating

The ignition quality of a fuel refers to how well it self-ignites under heat and pressure. Diesel fuel's ignition quality is measured by the cetane rating. To get a *cetane number rating*, a fuel is compared to cetane, a colorless, liquid hydrocarbon that has excellent ignition qualities. Cetane is rated at 100. The higher the cetane number, the shorter the ignition lag time (delay time) from the point the fuel enters the combustion chamber until it ignites. The exact rating is determined by mixing the cetane with a chemical called methyl-napthalene, which is rated at zero since it does not ignite. The percentage of cetane mixed with methyl-napthalene that produces a similar ignition quality to the fuel being tested is the cetane number rating. Ignition quality and flash point should not be confused.

Flash point is the lowest temperature at which the fuel burns when ignited by an external source.

The quality of gasoline is measured by octane, which indicates the resistance of a fuel to self-ignite (knock). Premium gasoline has poor ignition quality, since it burns slower than regular gasoline and has more resistance to preignition and detonation. The higher an octane number, the more resistance a fuel has to knocking. Diesel fuel cetane ratings are the opposite of gasoline octane ratings, figure 4-6. For automotive diesels, the recommended cetane rating is approximately 45, figure 4-7.

Carbon Residue

Carbon residue is the material left in the combustion chamber after burning. It is found not only in diesel

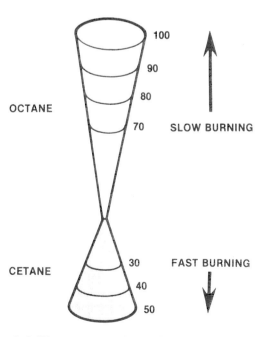

Figure 4-6 The octane scale is the opposite of the cetane scale. *(Courtesy of Ford Motor Company)*

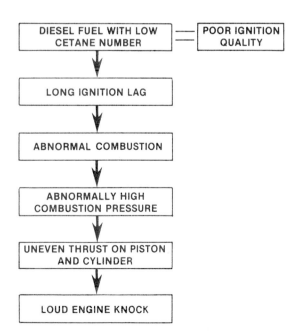

Figure 4-7 Note the effect of using the wrong cetane-rated fuel. *(Courtesy of Ford Motor Company)*

engines, but in other engines that burn hydrocarbon fuel.

The amount of carbon residue left by diesel fuel depends on the quality and the volatility of that fuel. Fuel that has a low volatility is much more prone to leaving carbon residue. The small, high-speed diesels found in automobiles require a high-quality fuel, since they cannot tolerate excessive carbon deposits. Large, low-speed industrial diesels are relatively unaffected by carbon deposits and can run on low-quality fuel.

Sulfur Content

Sulfur content is common in fuels made from low-quality crude oil. Refining the oil removes only a portion of the sulfur. Sulfur increases ring and cylinder wear, causing the formation of varnish on the piston skirts and sludge in the oil pan. Changing the oil frequently or switching fuels often helps prevent wear.

Fuels that have a high sulfur content are often high in various nitrogen compounds. These nitrogen compounds, like the sulfur, form corrosive chemicals causing excessive engine wear.

Water Content

Water in diesel fuel is a major problem because water and diesel fuel readily mix. Careless storage and distribution of diesel fuel invites problems. Diesel fuel that appears cloudy often contains water. Some of the problems that water causes are:

1. Corrosion of the fuel system. This can cause the fuel filter to plug with rust particles.
2. Icing of the fuel system. Wherever the water collects and the temperature is low enough, ice forms, causing severe damage to the fuel system components.
3. Inadequate lubrication of the injection pump and nozzles. Water does not have good lubricating qualities.
4. Bacteria growth in diesel fuel.

Bacteria Content

Diesel fuel is attacked by various fungi and bacteria. They ingest the diesel fuel as food, changing it to their waste products—a slimy, gelatin-type growth. This growth not only plugs the fuel system but also produces an acid that is corrosive to fuel system components. Because the fuel may contain harmful organisms, any wound exposed to diesel fuel should be cleaned immediately. Fungicides and bactericides, which prevent their formation and growth, are available.

COMMERCIAL FUEL RATINGS

There are three grades of diesel fuel for automotive use: 1-D, 2-D, and 4-D. At one time there was a grade 3-D, but it has been discontinued.

Grade 1-D is a kerosene-type fuel that has a lower viscosity, lower wax content, and lower BTU per gallon than grade 2-D. It is also more volatile than 2-D.

Grade 2-D is the fuel recommended for automotive and some industrial applications.

Grade 4-D is a fuel for low- and medium-speed engines, figure 4-8.

Heating fuel, which is similar to grade 2-D fuel, should not be used in automotive applications. Heating fuel does not meet the strict standards or have the needed additives for automotive use.

BLENDED FUELS

In cold climates it is often necessary to run on a blended fuel. A blended fuel reduces the WAP and pour point, allowing the fuel to flow at low temperatures. Typically, grade 1-D fuel is used to lower the WAP and pour point of grade 2-D fuel. Each manufacturer has specific instructions on what blend should be used at certain temperatures. Usually, a 10% increase of grade 1-D to grade 2-D lowers the WAP by 2°F (1°C). However, since grade 1-D has a lower heat energy content, fuel economy also decreases.

Additives are chemicals added at the refinery to lower the WAP and pour point. At the refinery, the composition of the oil and wax content is known. The proper additives are blended with the fuel to give it the desired properties. Additives used in the aftermarket by owners and technicians may or may not work because of variations in oil composition. Furthermore, use of additives may violate the manufacturer's warranty.

CAUTION: Never blend gasoline with diesel fuel.

Gasoline mixed with diesel fuel can create a powerful bomb. Diesel fuel alone in the tank emits very little vapor. Gasoline fills the tank with fumes that are too rich to burn. When mixed together, the combination of fuel vapor

	1-D	2-D	4-D
Minimum Flash Point, °F	100 or Legal	125 or Legal	130 or Legal
Viscosity, 100°F			
Minimum	1.4	2.0	5.8
Maximum	2.5	5.8	26.4
Carbon Residue, Weight Percent Maximum	0.15	0.35	—
Ash, Weight Percent Maximum	0.01	0.02	0.10
Sulphur, Weight Percent Maximum	0.50	1.0	2.0
Ignition Quality, Cetane No. Minimum	40	40	30
Distillation Temperature °F, 90% Evaporated			
Minimum	—	540	—
Maximum	550	576	—

Figure 4-8 Comparison of grades 1-D, 2-D, and 4-D diesel fuel *(Courtesy of Ford Motor Company)*

and air is potentially explosive. This mixture can be ignited in a variety of ways. A spark created by a static charge can occur merely by filling the tank. A person performing mechanical work on the vehicle can create a spark with tools or a lighted cigarette. Also, if the vehicle is in an accident, the fuel tank can explode.

FUEL STORAGE

Clean fuel for operating diesel engines is essential. Adequate containers are necessary to store fuel until it is used. Technicians who keep a small supply of diesel fuel on hand should be aware of a few facts:

1. Diesel fuel ages and will go stale. Keep a fresh supply available.
2. Variations in heat and humidity tend to create condensation in the fuel storage containers. Fuel containers should be kept where the temperature is relatively moderate and out of direct sunlight.
3. Never store diesel fuel in galvanized containers. Diesel fuel causes the galvanizing to flake off, contaminating the fuel system and clogging the fuel filters.
4. Containers should always be properly labeled and identified as containing diesel fuel.
5. Never add alcohol to diesel fuel. This lowers the flash point of the fuel.

SUMMARY

Diesel fuel has several characteristics different from gasoline. Diesel fuel has a higher heat content, specific gravity, and viscosity. Diesel fuel is more sensitive to cold weather. WAP (cloud point) is the temperature at which wax crystals appear. Pour point is the temperature at which diesel fuel solidifies and no longer flows. The cetane rating is the opposite of the octane rating. The higher the cetane number, the shorter the ignition delay time. Carbon residue is the material left after combustion. Small, high-speed diesels cannot tolerate excessive carbon deposits. Sulfur and nitrogen compounds create corrosive chemicals causing premature engine wear. Diesel fuel has an affinity for water. Care must be taken to keep the water content to a minimum.

Diesel fuel for automotive use comes in three grades: 1-D, 2-D, and 4-D. Grade 2-D is the recommended fuel for diesel engines in cars and trucks under most conditions. It is often blended with grade 1-D to lower the WAP and pour point.

Gasoline should never be mixed with diesel fuels. The combination of the fuels can create a powerful bomb when ignited by a spark.

Diesel fuel must be stored properly to prevent stale fuel and water contamination. It should never be stored in a galvanized container and never mixed with alcohol. Fuel containers should be clearly marked and identified.

CHAPTER 4 QUESTIONS

1. List five differences between diesel fuel and gasoline.
2. Explain the effects of cold weather on diesel fuel.
3. Explain how cold weather's effects on diesel fuel are compensated.
4. Explain the difference between cetane and octane ratings.
5. Explain how ignition lag time is influenced by the cetane rating.
6. Explain why diesel fuel and gasoline should never be mixed.
7. Explain why a wound contaminated with diesel fuel should be treated immediately.
8. List five rules for storing diesel fuel.

chapter 5
Basic Diesel Fuel Systems

Objectives

In this chapter you will learn:

- The components of the diesel fuel system
- How fuel flows through the diesel fuel system
- Two types of fuel lines
- Two locations for water-in-fuel sensors
- How a water-in-fuel separator works
- Three types of fuel supply pumps
- The operation of the fuel heater
- Five functions of the fuel injection pump
- Two functions of the nozzle
- Basic fuel system services

COMPONENTS OF THE DIESEL FUEL SYSTEM

The components of the diesel fuel system are the fuel tank and pick-up unit, fuel lines, water-in-fuel sensor, water-in-fuel separator (not found on all makes), fuel filter, fuel supply pump (not found externally on all makes), fuel heater, fuel injection pump, and injection nozzles.

Fuel Flow through the Diesel Fuel System

Fuel stored in the fuel tank is drawn through the pick-up unit by a fuel lift pump, figure 5-1. Fuel travels through a water-in-fuel separator (optional) to the fuel supply (lift) pump. From there it goes through a fuel heater (optional), then through a filter to the fuel injection pump. The fuel injection pump pressurizes the fuel to a very high pressure and sends it to the nozzle at the proper time. The nozzle atomizes and sprays the fuel into the combustion chamber. Excess fuel pumped by the fuel lift pump cools and lubricates the injection pump and nozzles. This fuel, called return fuel, is sent back to the fuel tank through the fuel return line. The fuel system also prevents any air from entering the system, since air causes rough running.

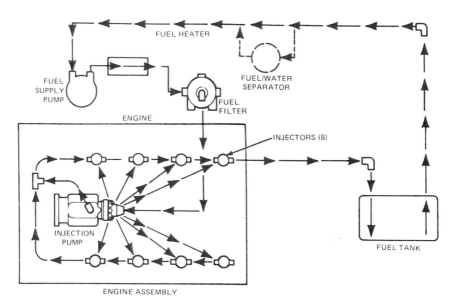

Figure 5-1 Fuel flow through a typical system (not representative of all fuel systems in use) *(Courtesy of Ford Motor Company)*

Fuel Tank and Pick-up Unit

The fuel tank on any vehicle obviously stores fuel. The fuel tank in a diesel-powered vehicle is usually larger than in a gasoline-powered vehicle. This allows the vehicle to travel a greater range. The greater range is needed because filling stations that handle diesel fuel are not as plentiful as gasoline service stations. Some tanks are designed with a small reservoir section. This prevents air from entering the system when the tank is low on fuel and when the vehicle makes a sudden change in movement or a wide turn.

The fuel pick-up is similar to the conventional gasoline pick-up, containing a fuel-level sending unit and a strainer. However, different features are added to accommodate the characteristics of diesel fuel and the diesel fuel system.

One type of pick-up unit uses a fuel pick-up filter (sock) and check valve assembly, figure 5-2. The sock is fastened to a pick-up tube a short distance above the bottom of the tank. The sock strains large particles, limits entry of water, and draws fuel from the bottom up to the pick-up tube, acting as a wick. A check valve is added to the sock. It is designed to open when the sock is restricted by wax or ice, allowing the engine to run. This type of pick-up unit is usually equipped with a water-in-fuel sensor since the sock holds back only a certain amount of water before it passes and this system may

not have a water-in-fuel separator. Obviously, it is important that the owner know immediately if there is an excessive amount of water in the tank requiring purging of the fuel system. The return pipe returns excess fuel near the pick-up filter. There are two basic reasons for this. First, diesel fuel, when sprayed into the air, oxidizes, changing into a tarlike substance that causes the injection pump and nozzles to clog or stick. Allowing return fuel to enter low in the tank, where it will almost always be submerged by diesel fuel, prevents this. Second, the return fuel, having been heated by the fuel system and engine, melts ice and wax crystals plugging the fuel intake.

Another type of pick-up unit places the fuel intake in contact with the bottom of the tank supported by several small nibs, figure 5-3. Fuel going past the nibs accelerates, creating a pressure drop. This pressure drop attracts water. Water in the diesel fuel is drawn to the intake, accelerated, and forced through the nylon screen along with the fuel. Using this type of pick-up unit necessitates a water-in-fuel separator to filter and eliminate the water. This pick-up unit also has low placement of the return line and a bypass valve in the event the intake is clogged with wax or ice, figure 5-4.

With either pick-up unit, the tank must be at least one-quarter full for the bypass valve to pick up fuel.

Fuel Lines

Fuel lines are for carrying fuel without leaking or admitting air. They can be divided into two groups—low pressure and high pressure.

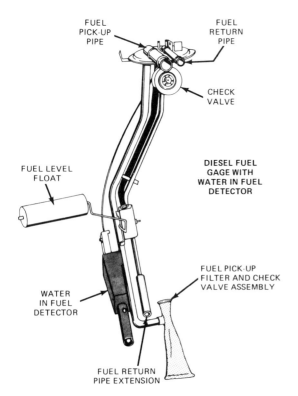

Figure 5-2 Diesel fuel pick-up unit with water-in-fuel detector (Courtesy of Oldsmobile Division)

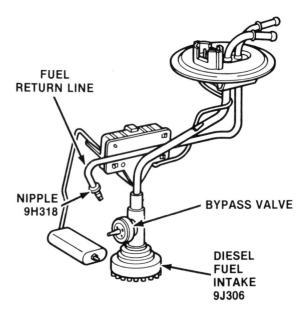

Figure 5-3 Fuel pick-up unit (Courtesy of Ford Motor Company)

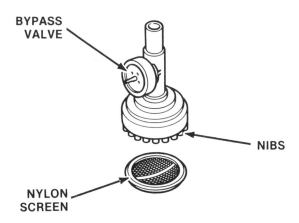

Figure 5-4 Fuel intake (*Courtesy of Ford Motor Company*)

Low-pressure lines are the lines that carry fuel to the injection pump and are used to return excess fuel from the injection pump and injection nozzles back to the tank, figure 5-5. Diesel lines have a slightly larger inside diameter than gasoline-powered models because diesel fuel is slightly more viscous and tends to thicken in cold weather. Fuel pressure here, under normal circumstances, rarely exceeds 12 psi (82.74 kPa). These lines can be made of plastic, synthetic rubber, or steel, all of which must be compatible with diesel fuel. Synthetic rubber and plastic lines used on gasoline engines may not be compatible with diesel fuel.

High-pressure lines carry the pressurized fuel from the injection pump to the injection nozzles. Here, line pressure can go to as high as 6000 psi (41,370 kPa). Special steels are used to resist the high pressures and corrosion that might form within the line. All lines are of equal length and inside diameter; therefore, each line holds the same amount of fuel. This assures that each nozzle will receive the proper amount of fuel at the proper time.

Any restriction within the fuel system will result in poor engine performance. Both high- and low-pressure fuel lines should never be dented or bent.

Water-in-Fuel Sensor and Separator

The water-in-fuel (WIF) sensor is located in one of two places: the fuel tank or the water-in-fuel separator. If it is located in the fuel tank, the sensor is mounted on the sending unit pick-up tube. When water reaches the sensor, it provides a ground for the WIF warning light on the dash. If the light comes on, the owner should immediately have the fuel system purged.

To ensure that the WIF circuit is working properly, the WIF light comes on and remains on for a few seconds each time the vehicle is started. After a 15-20 second delay, if water is present, the WIF light comes on and stays on. The sensor detects from 1½ to 2½ gallons (22–37.8 L) of water.

The WIF separator (sometimes called a sedimentor) is usually mounted between the fuel tank and filter, figure 5-6. Its purpose is to remove water from the fuel and retain it until the water can be drained. The WIF accomplishes this in two ways. First, water, being heavier

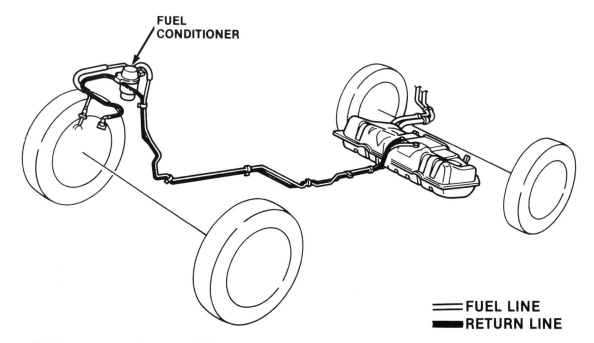

Figure 5-5 Low-pressure lines carry fuel to and from the injection pump. (*Courtesy of Ford Motor Company*)

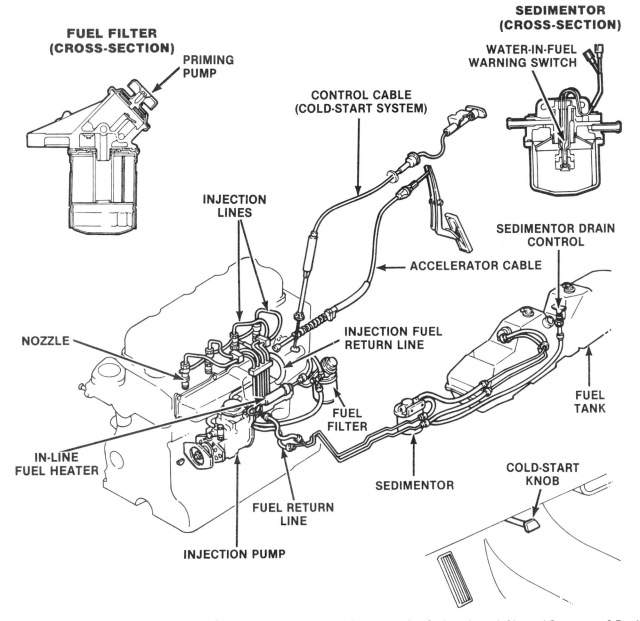

Figure 5-6 The sedimentor (water-in-fuel separator) is placed between the fuel tank and filter. (*Courtesy of Ford Motor Company*)

than diesel fuel, tends to collect at the bottom of separator. Second, a special sock or filter is used to attract water. Water clings to the sock and collects at the bottom of the separator, figure 5-7.

A sensor located in the separator comes on when the water reaches a predetermined level. A float or detector is often used to indicate water level. A float that is lighter than water but heavier than diesel fuel rises as the water level rises. At the appropriate point, a magnet inside the float closes a set of contacts, figure 5-8. When the contacts are closed, the circuit is complete, allowing a warning light or buzzer to come on. A detector works on the principle that water is a better conductor than diesel fuel

and provides a ground for the warning lamp when water is present, figure 5-9.

When the WIF light or buzzer comes on, the owner must have the vehicle serviced immediately to prevent costly repairs. Servicing the fuel system involves purging the entire fuel system and installing a new filter, or simply draining the water out of the separator.

Fuel Filter(s)

The purpose of any filter is to trap and prevent contaminants from reaching a critical amount. Particles larger than 10 microns can damage the injection pump,

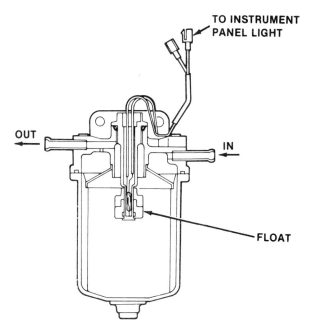

Figure 5-7 Cross section of a water-in-fuel separator (*Courtesy of Ford Motor Company*)

figure 5-10. In the diesel fuel system, the fuel filter prevents contaminants from reaching the finely machined, extremely close-tolerance fuel injection pump components.

There may be one or two filters in the system other than the filter (strainer) in the fuel tank. If two filters are used, the first filter is called the primary filter. Its purpose is to trap large particles of dirt (80 microns) and water. There may be a drain mounted on this filter to remove water. The next filter is the secondary filter. Its

purpose is to trap the smaller particles missed by the primary filter.

If only one filter is used, it is generally of a two-stage design. The first stage, like the primary filter, traps larger particles of dirt and sediment. The second stage traps the smaller particles, figure 5-11. Water will be trapped in the filter but will eventually be drained.

Some filter housings are equipped with priming pumps that operate by hand, figure 5-12. This aids in forcing fuel through the system, purging it of air.

Fuel Supply (Lift) Pumps

The fuel supply pump must draw an adequate amount of fuel from the tank and send it, under pressure, to the fuel injection pump. Fuel supply pumps pump more fuel than the engine can burn. The excess fuel is used for cooling and lubricating the injection pump and nozzles. The supply pump is located inside or outside the fuel injection pump. We will deal only with supply pumps mounted externally. Supply pumps mounted internally will be discussed in a later chapter.

Many fuel systems require an external fuel supply pump to draw the fuel out of the tank and send it, under low pressure, to the injection pump. Supply pumps can be divided into two categories—mechanical and electrical.

Mechanical supply pumps are either the plunger type or diaphragm type. The plunger-type supply pump is mounted on the injection pump housing. The plunger is actuated by an eccentric on the injection pump camshaft.

As the plunger moves toward the camshaft, fuel is drawn in through the inlet (suction) check valve, filling the chamber above the plunger, figure 5-13a.

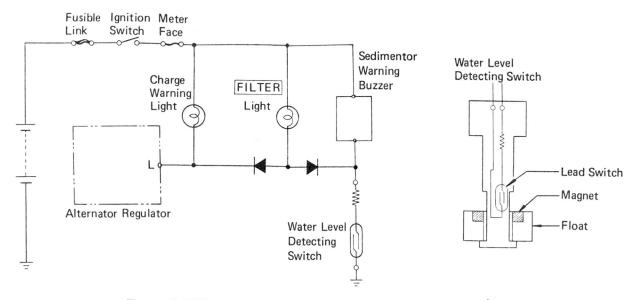

Figure 5-8 Water-in-fuel circuit and water level detecting switch.

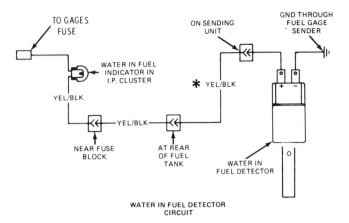

Figure 5-9 Water detectors may be located in the fuel tank or separator. Water level is sensed by a capacitive probe. When the water reaches a predetermined level, the detector switches the warning lamp on. (*Courtesy of Oldsmobile Division*)

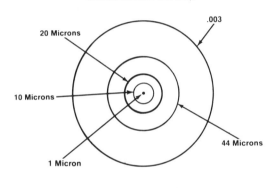

Figure 5-10 A comparison of the relative size of micron particles (*Courtesy of General Motors Product Service Training*)

This unit filters the fuel, separates water from the fuel, and warms the fuel when it is cold. Sensors dectect water level and filter restriction. A petcock is provided to drain the water.

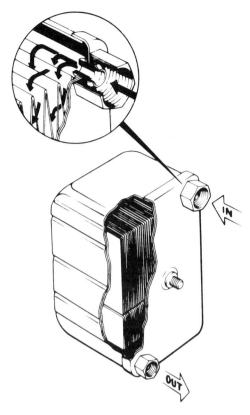

Figure 5-11 A fuel filter with a dual-stage design. (*Courtesy of General Motors Product Service Training*)

FUEL FILTER

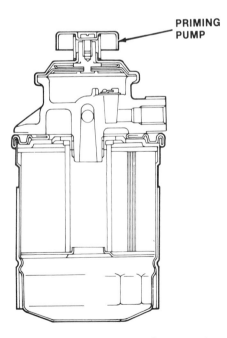

Figure 5-12 A hand priming pump aids in purging air from the fuel system. (*Courtesy of Ford Motor Company*)

When the eccentric forces the plunger away from the camshaft, the outlet (pressure) check valve is opened, the inlet side is closed, and the fuel above the plunger is pressurized, figure 5-13b. Further rotation of the eccentric allows the plunger to travel back. The plunger spring, which is compressed, pushes the plunger back, thereby pressurizing fuel behind the plunger.

When the injection pump cannot use all the fuel supplied by the supply pump, pressure is maintained behind the

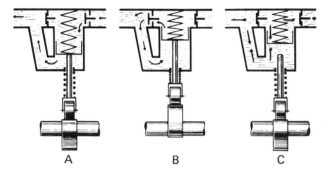

Figure 5-13a. Inlet valve open, outlet valve closed, pump is spring actuated (plunger). **Figure 5-13**b. Inlet valve closed, outlet valve open, pump is cam actuated. **Figure 5-13**c. Inlet valve closed, outlet valve closed, self-regulating. *(Courtesy of United Technologies Diesel Systems)*

plunger and both check valves are closed, figure 5-13c. This prevents the plunger from traveling back to its starting point and from drawing in more fuel as the eccentric rotates. As fuel is used by the injection pump, the plunger recedes to its starting position, figure 5-13a.

The fuel flow in the diaphragm type is the same as those found on gasoline engines. However, on some engines an eccentric is mounted on the crankshaft instead of the camshaft. The reason is that the injection pump drive occupies the area normally used for the fuel pump. Another point to remember is that a gasoline diaphragm pump is not interchangeable with one from a diesel engine because the internal parts of the fuel pump are compatible with diesel fuel only.

An electric or solenoid supply pump may be used when it is impractical to use a mechanical fuel pump. This type can be mounted in a variety of locations, and there are no rubber parts to wear out. With the ignition key turned on, the transistor inside the pump is turned on. This energizes the primary winding, pulling the plunger down into the winding, and opening the inlet valve. As the clearance volume increases, pressure decreases, and fuel rushes in to fill the volume displaced by the plunger, figure 5-14. With the plunger all the way down, the pressure spring is compressed, and the plunger acts as would an iron core on a solenoid. This induces a voltage

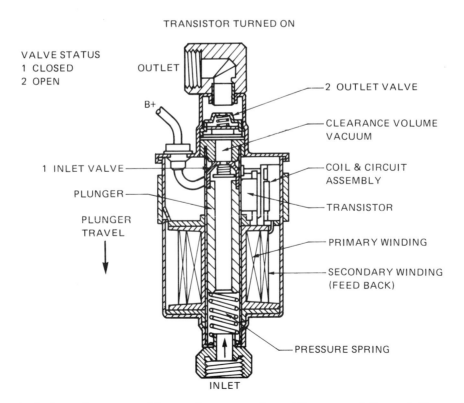

Figure 5-14 Electric fuel supply pump with transistor turned on *(Courtesy of General Motors Product Service Training)*

TRANSISTOR TURNED OFF

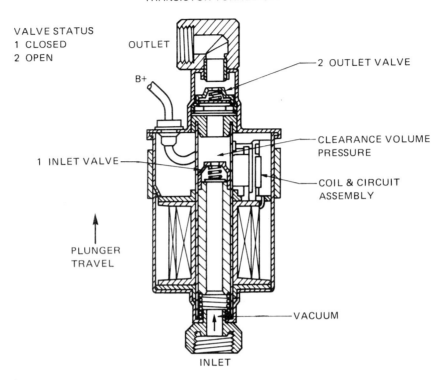

VALVE STATUS
1 CLOSED
2 OPEN

OUTLET

B+

2 OUTLET VALVE

CLEARANCE VOLUME PRESSURE

1 INLET VALVE

COIL & CIRCUIT ASSEMBLY

PLUNGER TRAVEL

VACUUM

INLET

Figure 5-15 Electric fuel supply pump with transistor turned off (*Courtesy of General Motors Product Service Training*)

in the secondary winding that causes the transistor to turn off current to the primary winding. The pressure spring pushes the plunger up, closing the inlet valve, opening the outlet valve, and delivering pressurized fuel, figure 5-15. When the plunger returns to the top, the cycle is repeated. Fuel cools the solenoid windings. Hence, it is important that the pump has enough fuel to prevent the windings from overheating.

Fuel Heaters

Automobiles and small trucks use an electric fuel heater mounted between the tank and injection pump. Its job is to raise the temperature of the fuel when the temperature is cold enough to form wax crystals. This prevents the wax crystals from plugging the fuel filter when the engine is first started. Once the engine is started, engine heat and warm return fuel raise the fuel's temperature. The fuel heater may be part of the fuel line or fuel conditioner.

The in-line heater, figure 5-16, is thermostatically controlled, using a bimetal switch, which turns on at 20°F (−13°C) and shuts off at 50°F (10°C). When the bimetal switch is closed, power flows through a thermal fuse and breakaway solder joint. These serve to protect the system from overheating because of a dry line or no fuel flow.

Figure 5-16 Pointer is resting on in-line fuel heater.

From the solder joint, current flows through the resistance heating element to ground, figure 5-17. The heating element is insulated to prevent heat loss.

The fuel heater, located in the fuel conditioner, works essentially the same way, with bimetallic switches sending current to the heater when the fuel temperature is from 30°F (−1°C) to 55°F (13°C). There are two self-regulating heaters within the conditioner, figure 5-18. As heater temperature increases so does electrical resistance,

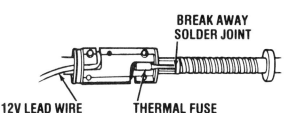

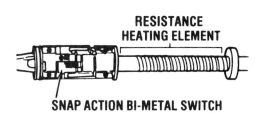

Figure 5-17 Cutaway of in-line fuel heater (Courtesy of General Motors Product Service Training)

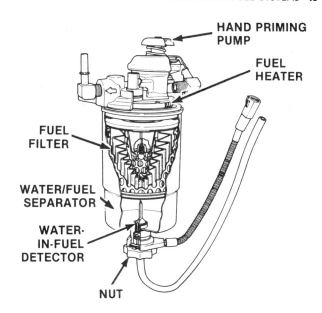

limiting current flow and maximum temperature. This type of heater, called a positive temperature coefficient heater, eliminates the need for a control module or limiting device.

As previously mentioned, for the fuel heater to be effective, modifications to the fuel pick-up unit were needed. A bypass valve located above the intake opens when the intake is clogged with wax. Since the bypass valve is located above the intake, the owner must keep the tank at least one-quarter full to allow fuel to be drawn in. There may also be a check valve on the return fuel line located at the top of the pick-up unit. This check valve opens when the lower portion of the return line is plugged, allowing fuel to flow.

Fuel Injection Pump

The fuel injection pump performs several tasks. These include metering the fuel, pressurizing the fuel, timing the fuel delivery, governing engine speed, and shutting off the engine.

Fuel injection pumps meter, or control, the amount of fuel the cylinder receives. Varying the amount of fuel controls engine power and speed. Fuel injection pump manufacturers have different methods for accomplishing this task, which will be discussed in later chapters.

Fuel must be delivered to the nozzles under high pressure for the nozzles and engine to run smoothly. The fuel injection pump may contain one or more pumping element.

A pumping element is made of the components necessary to pressurize only the fuel. Arrangement and number of pumping elements vary according to need and the manufacturer. It is at the pumping element that low-pressure fuel is pressurized to as much as 6000 psi

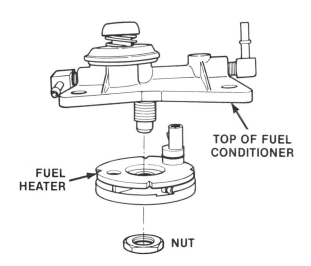

Figure 5-18 Fuel conditioner and heater. (Courtesy of Ford Motor Company)

(41,370 kPa). Pumping elements must be made with complete precision and quality materials to do the job properly.

A critical point in engine performance is when the fuel gets to the cylinder. As engine speed increases, the fuel must be injected earlier, to have enough time to mix and burn; achieving most out of the fuel's heat energy. The injection pump automatically advances fuel injection timing.

Fuel injection pumps use governors for load and speed control under various conditions. Governors used in cars and small trucks are the *minimum-maximum* design. That is, the governor controls the idle RPM and top engine RPM. Engine RPM between idle and top engine speed is controlled directly by the driver. This differs from governors used on tractors and generators. On these vehicles, the governor is of the load-sensing type. With the throttle set in one position, the governor varies the amount of fuel, maintaining a constant speed as the load changes. On a car, the driver must move the accelerator pedal to maintain a constant speed as different loads are encountered. Again, only idle and top engine speeds are governed.

A fuel shut-off mechanism is used to prevent fuel from flowing to the pumping elements. This prevents the engine from running. On cars and small trucks, the fuel shut-off mechanism is activated by turning the key on.

There are a variety of fuel pump configurations, but only two types are used in automobiles and small trucks—distributor and in-line.

The *distributor-type* configuration is the most common for small, high-speed diesels. It is relatively inexpensive, compact, and lightweight. The distributor-type injection pump uses one pumping element to supply fuel to all the engine's cylinders, figure 5-19. From one pumping element fuel is distributed to the appropriate cylinder, according to the engine's firing order.

The *in-line* pump configuration uses one pumping element for each engine cylinder, Figure 5-20. If there are five engine cylinders, then the fuel injection pump has five pumping elements. This type is the least common design in cars and small trucks though it has long been in use and is very common on larger diesel engines. It is very dependable, but more expensive than the distributor type.

Regardless of the type of injection pump, it is a complex component manufactured to the strictest tolerances, using only quality materials. Contaminated fuel and misuse can quickly destroy the close-fitting precision parts. Service of these pumps is discussed in the appropriate chapters.

Nozzle Holders and Nozzles

Nozzle holders and nozzles are located in the cylinder head with the nozzle end facing the combustion chamber. The nozzle holder contains and secures the nozzle. Fuel lines are also connected to the holder. The nozzle contains the parts that direct fuel sent by the injection pump under high pressure into the combustion chamber, figure 5-21. When injecting the fuel into the combustion chamber, the nozzle (a) atomizes the fuel, and (b) spreads the fuel spray in a particular pattern.

When the fuel is atomized it is broken down into tiny droplets that allow easy mixing with the air. To ensure

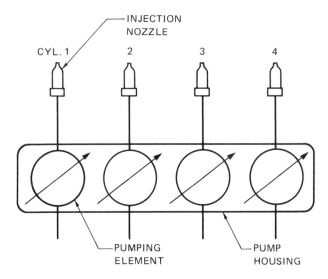

Figure 5-20 An in-line fuel system uses one pumping element for each injection nozzle.

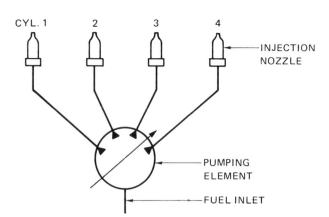

Figure 5-19 A distributor fuel system uses one pumping element for all injection nozzles.

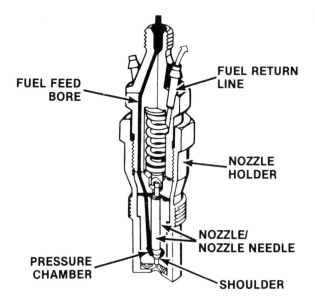

Figure 5-21 Injection nozzle *(Courtesy of Ford Motor Company)*

that all the fuel is mixed with air, the nozzle spreads the fuel out in a pattern. This is called the spray pattern and is tailored for each combustion chamber design. Nozzles and nozzle holders come in many designs and have different modes of operation. This will be discussed in Chapter 9.

BASIC FUEL SYSTEM SERVICE

The integrity of the diesel fuel system must be maintained. Any air or dirt that enters can wreak havoc on the fuel system. All fuel lines must be free of distortion and all connections tight. If any of these components are not doing their job properly, the result can be a poor-running diesel and expensive repairs.

When servicing any component on a diesel fuel system, extra care is needed to prevent dirt and air from entering the system. Clean the area you are working in. Cap all exposed openings to the fuel system.

Replacing a Fuel Filter

Fuel filters should be replaced periodically as specified by the manufacturer. This requires the following:

- new filter
- cleaning cloths or towels
- appropriate hand tools
- manufacturer's service manual

Replace the fuel filter by following these procedures:

1. Clean the filter area before removing any lines.
2. Disconnect fuel filter and cap all exposed openings.
3. Fill fuel filter with diesel fuel. If this is not possible, go to step 4.
4. Install fuel filter. If it was impossible to fill the filter with fuel, leave the outlet fitting slightly loose. Crank the engine until fuel appears at the outlet fitting and tighten the fitting. On some fuel systems, air can be bled off by pumping the primer pump several times.
5. Wipe off all spilled fuel and dry all connections.
6. Start the engine and let it run for a few minutes to clear the system of any air that may have leaked in; inspect for any fuel leaks.
7. If the filter suddenly became plugged without warning, it is important to find out why, before giving the vehicle back to the owner. If dirt and sediment plugged the filter, poor quality fuel was the cause of the problem. In cold climates, water and ice are often to blame. To determine which is the cause, put the filter in a pan and let it stand at room temperature. If fuel flows out, wax caused the problem and the WAP of fuel must be lowered. If

water flows out, the rest of the fuel system must be purged.

Replacing a High-Pressure Fuel Line

This task requires the following:

- the EXACT fuel line that is to be replaced
- appropriate hand tools
- manufacturer's service manual

Replace the fuel line by following these procedures:

1. Disconnect the fuel line brackets.
2. Disconnect fuel line and cap all exposed openings.
3. Install new fuel line and connect fuel line brackets.

CAUTION: Do not distort the line. This will result in poor engine performance.

---

4. Tighten to specifications the line fitting at the injection pump. Leave the line fitting at the nozzle slightly loose.
5. Start the engine. When fuel appears at the fitting, tighten the fitting to specifications. Inspect for leakage.

Removing Water from the Fuel System

The procedure for removing water from the fuel system will vary slightly according to the amount of water in the system and the vehicle manufacturer. The following are two procedures for removing water.

When water has collected in the WIF separator over a period of time, simply open the drain and allow the water to run out until fuel appears. Dry off any components that may have fuel spilled on them. Start the engine and inspect for leaks.

Large amounts of water suddenly added to the fuel tank are the result of poor storage. The owner will complain that the WIF indicator came on just after filling up. When this condition occurs, the fuel tank must be drained and the fuel system purged. Siphoning or pumping the fuel tank is the most convenient method.

This task requires the following:

- a siphon or pump
- a five-gallon container
- a new fuel filter
- approximately 8 feet of fuel line
- manufacturer's service manual

Remove water from the fuel system by following these procedures:

1. Siphon or pump out the fuel tank according to the manufacturer's instructions. If recommended, remove the tank for further draining and inspection.
2. Fill the tank with fresh fuel.
3. Drain the water separator if equipped.
4. Disconnect the fuel line before the filter and attach a hose to the fuel line opening. Put the other end of the fuel line into a container.
5. Disconnect the fuel shut-off solenoid and crank the engine until clean fuel comes out. Do not crank for more than 30 seconds at a time to prevent starter motor damage.
6. Install a new fuel filter. Leave the line fitting going to the pump, loose.
7. Disconnect the fuel return line on the pump, and connect a fuel line with the other end running into a container.
8. Connect the fuel shut-off solenoid.
9. Crack open all high-pressure lines at the nozzles.
10. Crank the engine until fuel appears at the line fitting on the filter. Tighten the line fitting.
11. Crank the engine with accelerator held to the floor until uncontaminated fuel appears in the container. Do not crank for more than 30 seconds at a time and allow 2-minute intervals.
12. Tighten the lines at the nozzles.
13. Start the engine and run it for 15 minutes. Check to be sure the container does not overflow.
14. Connect the fuel injection pump return line and inspect for leaks.

If the fuel injection pump is suspected of damage, consult the appropriate chapter or manufacturer's service manual on inspection.

SUMMARY

Fuel is drawn out of a fuel tank through the pick-up unit, WIF separator (optional), or primary filter (optional) to an external lift pump, if equipped. From there the fuel flows to the fuel heater (optional), to a filter, and finally to the fuel injection pump. The fuel injection pump sends pressurized fuel to the nozzles, which spray the fuel into the combustion chamber. Excess fuel is used to cool and lubricate the injection pump and nozzles.

The fuel tank and pick-up unit are similar to their counterparts on gasoline-powered vehicles. Diesel fuel tanks are larger and have a small reservoir area to prevent air from entering the system.

Fuel lines are divided into two categories—low pressure and high pressure.

Low-pressure lines must be compatible with diesel fuel. High-pressure lines have the same length and internal diameter to ensure that each cylinder receives the proper amount of fuel at the proper time. Under NO circumstances should the fuel lines be kinked or bent.

WIF sensors may be located in the tank or WIF separator, if equipped. WIF separators are located between the fuel injection pump and fuel tank. They are designed to separate water from diesel fuel and provide a means to drain the water when the separator is full.

Fuel supply pumps pump fuel from the tank to the injection pump. Additional fuel, more than is needed for combustion, is pumped to cool and lubricate the injection pump and nozzles. There are three types of external fuel supply pumps—the plunger type, diaphragm type, and electric (solenoid) type.

Five functions of the fuel injection pump are metering the fuel, pressurizing the fuel, timing the fuel delivery, governing engine speed, and shutting off the engine.

The fuel injection pump is a precision component that must carry out the mentioned tasks for proper engine performance.

Two functions of the nozzle are (1) atomizing the fuel, and (2) spreading the fuel in a particular pattern.

Whenever servicing the fuel system, the integrity of the system must be maintained. Care is needed to prevent dirt and air from entering the system. Fuel lines must be of the proper size and must never be bent.

CHAPTER 5 QUESTIONS

1. List the components of the diesel fuel system.
2. With the aid of a diagram, trace fuel flow.
3. Describe the operation of the fuel pick-up units and how they differ.
4. List two types of fuel lines and where they are located.
5. Why are the inside diameter and length of a high-pressure fuel line important?
6. Name two locations for water-in-fuel sensors.
7. Explain how a water separator works.

8. Name three types of fuel supply pumps and describe how they work.
9. Describe the operation of the two types of fuel heaters.
10. List five functions of the fuel injection pump.
11. Name the speeds at which the minimum-maximum governor controls fuel.
12. Describe the difference between a minimum-maximum governor and a load-sensing governor.
13. Name two functions of the nozzle.

chapter 6
The Roosa Master DB2
Fuel Injection Pump

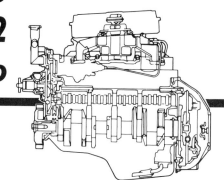

Objectives

In this chapter you will learn:
- To identify the Roosa Master DB2 injection pump
- To identify its main components
- The pattern of the fuel flow
- The operation of the transfer pump
- The operation of the regulator assembly
- How changes in fuel viscosity are compensated
- The charging and discharging cycle
- The operation of the delivery valve
- How fuel flows through the return oil circuit
- The operation of the minimum-maximum governor
- The operation of the automatic advance mechanism
- The operation of HPCA
- The operation of electric shut-off
- How to perform specified services on the injection pump

IDENTIFYING THE ROOSA MASTER DB2 INJECTION PUMP

To identify the Roosa Master DB2 injection pump, it is sometimes necessary to identify the model number system. The model number code is located on the side of the injection pump housing and is broken down as follows:

DB2 8 29 2N 3989

a. DB2:D—series pump, B—rotor, 2—2nd generation
b. 8—Number of cylinders served
c. 29—Abbreviation of plunger diameter
  29 =.290 in. (7.36 mm)
d. ZN—accessory code

This code pertains to combinations of special accessories such as electrical shut-off, automatic advance, and so on.

e. 3989—Specification number

This number determines selection of parts and adjustment for a given application. It must be included in any reference to the pump.

MAIN COMPONENTS

To understand the basic operating principles of the Roosa Master DB2 pump, it is necessary to become familiar with the main components. See and refer to figures 6-1 and 6-2 for the location of the main components.

The main rotating components are the drive shaft (1), transfer pump blades (5), distributor rotor (2), and governor (11).

The drive shaft engages the distributor rotor that is mounted in the hydraulic head. Where the drive shaft engages the rotor are two pumping plungers.

The plungers are pushed toward each other simultaneously by an internal cam ring through rollers and shoes that are carried in slots at the drive end of the rotor. The number of cam lobes equals the number of cylinders, figure 6-3.

The transfer pump at the rear of the rotor is of the positive displacement vane-type and is enclosed in the end cap. The end cap also houses the fuel inlet strainer and transfer pump pressure regulator. The face of the regulator assembly is compressed against the liner and distributor rotor and forms an end seal for the transfer pump. The injection pump is designed so that end thrust is against the face of the transfer pump pressure regulator.

The distributor rotor incorporates two charging ports and a single axial bore with one discharge port to service all discharge outlets to the injection lines, figure 6-4.

The hydraulic head contains the bore in which the rotor revolves, the metering valve bore, the charging ports, and the head discharge fittings. The number of charging ports and discharge fittings equals the number of cylinders.

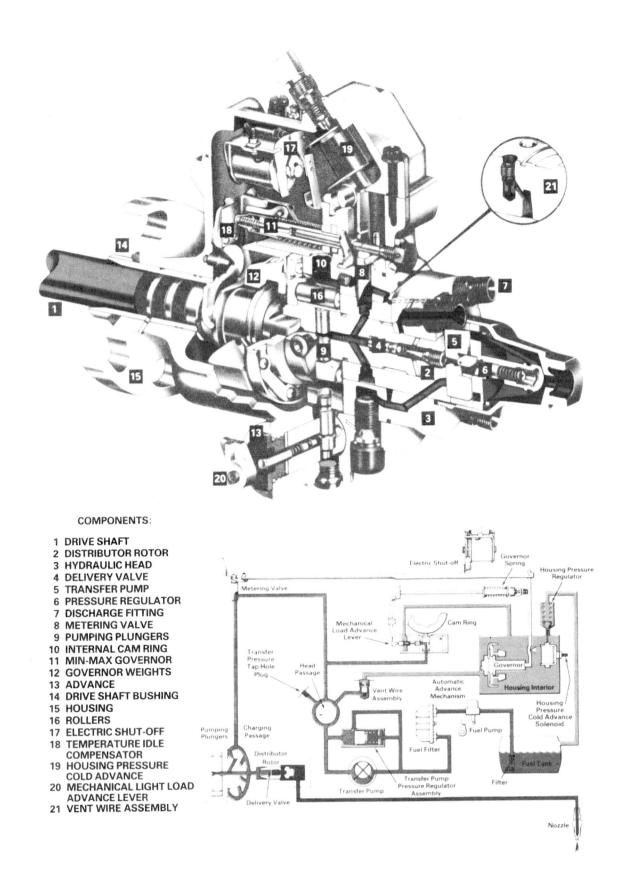

COMPONENTS:

1 DRIVE SHAFT
2 DISTRIBUTOR ROTOR
3 HYDRAULIC HEAD
4 DELIVERY VALVE
5 TRANSFER PUMP
6 PRESSURE REGULATOR
7 DISCHARGE FITTING
8 METERING VALVE
9 PUMPING PLUNGERS
10 INTERNAL CAM RING
11 MIN-MAX GOVERNOR
12 GOVERNOR WEIGHTS
13 ADVANCE
14 DRIVE SHAFT BUSHING
15 HOUSING
16 ROLLERS
17 ELECTRIC SHUT-OFF
18 TEMPERATURE IDLE
   COMPENSATOR
19 HOUSING PRESSURE
   COLD ADVANCE
20 MECHANICAL LIGHT LOAD
   ADVANCE LEVER
21 VENT WIRE ASSEMBLY

Figure 6-1 Cutaway of DB2 fuel injection pump with fuel flow schematic. (*Courtesy of General Motors Product Service Training*)

FUEL DISTRIBUTION

CHARGING

DISCHARGING

RETURN OIL CIRCUIT

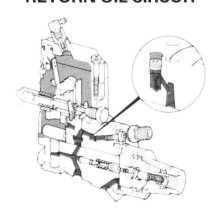

AUTOMATIC ADVANCE

REGULATOR ASSEMBLY

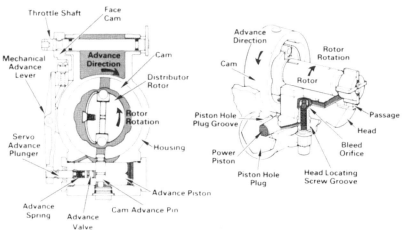

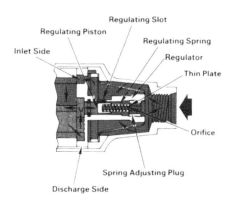

MECHANICAL GOVERNOR

TRANSFER PUMP

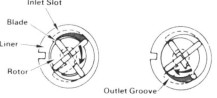

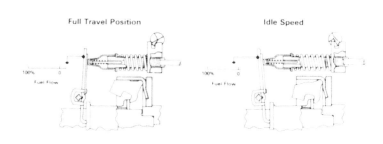

DELIVERY VALVE FUNCTION

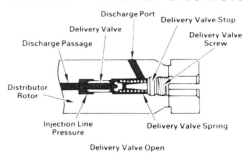

Figure 6-2 (*Courtesy of General Motors Product Service Training*)

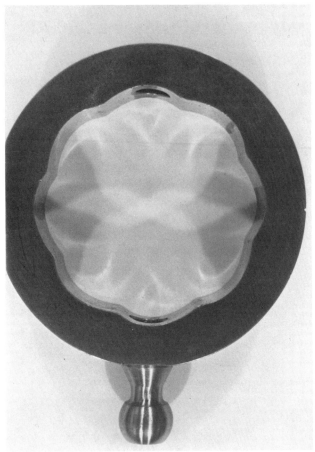

Figure 6-3 Cam ring with eight internal lobes and advance pin

In an eight-cylinder engine, the injection pump would have eight charging ports and eight discharge ports. The high-pressure injection lines to the nozzle are fastened to these discharge fittings.

The DB2 pump has its own mechanical "min-max" governor that rotates on the drive shaft.

The automatic advance is a hydraulic mechanism that advances or retards the pumping cycle.

FUEL FLOW

The operating principles of the pump can be understood more readily by following the fuel circuit during a complete pump cycle; see figures 6-1 and 6-2.

Fuel is drawn through a strainer in the tank by the life pump. Fuel at approximately 5½ to 6½ psi (37.92–44.81 kPa) flows through the filter into the pump inlet through the inlet screen (figure 6-1, 1). The fuel is also being drawn in by the transfer pump (2). Some fuel is bypassed through the pressure regulator assembly (3) to the suction side.

Fuel under transfer pump pressure flows through the center of the transfer pump rotor, past the rotor retainers

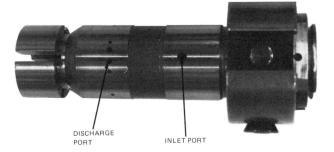

Figure 6-4 Distributor rotor

(4), into a circular groove on the rotor. It then flows through a connecting passage (5) in the head, to the automatic advance (6), up through a radial passage (7), and through a connecting passage (8), to the metering valve. The radial position of the metering valve, controlled by the governor at minimum and maximum engine speeds and by the driver at all points in between, regulates flow of the fuel into the radial charging passage (9), which incorporates the head charging ports.

As the rotor revolves, the two rotor inlet passages (10) register with the charging ports in the hydraulic head, allowing it to flow into the pumping chamber. With further rotation, the inlet passages move out of registry and the discharge port of the rotor registers with one of the head outlets. While the discharge port is open, the rollers (11) contact the cam lobes, forcing the plungers together. Fuel trapped between the plungers is then pressurized and delivered to the injection nozzle, then into the combustion chamber.

Self-lubrication of the pump is an inherent feature of the Roosa Master design.

An air vent passage (12) in the hydraulic head connects the outlet side of the transfer pump with the pump housing. This allows air and some fuel to bleed back to the fuel tank via the return line. The fuel thus bypassed fills the housing, lubricates internal components, cools, and carries off any small air bubbles. The pump operates with the housing completely full of fuel; there are no dead air spaces anywhere within the pump.

Operation and fuel flow within each of the major components are discussed in the following sections.

TRANSFER PUMP

The fuel transfer pump is a positive displacement pump consisting of a stationary liner and spring-loaded blades that are carried in slots in the rotor. Since the inside diameter of the liner is eccentric to the rotor axis, rotation causes the blades to move in the rotor slots. This blade movement changes the volume between the blade segments.

Transfer pump output volume and pressure increase as pump speed increases. Since displacement and pressure of the transfer pump can exceed injection requirements, some of the fuel is recirculated by means of the transfer pump regulator back to the inlet side of the transfer pump.

Figure 6-5 illustrates the pumping principle. Radial movement causes a volume increase in the quadrant between blades 1 and 2, figure 6-5a. In this position, the quadrant is in registry with the kidney-shaped inlet slot in the top portion of the regulator assembly. The increasing volume causes fuel to be pulled through the inlet fitting and filter screen into the transfer pump liner. Volume between the two blades continues to increase until blade 2 passes out of registry with the regulating slot. At this point, the rotor has reached a position where outward movement of blades 1 and 2 is negligible and volume is not changing, figure 6-5b. The fuel between the blades is being carried to the bottom of the transfer pump liner.

As blade 1 passes the edge of the kidney-shaped groove in the lower portion of the regulator assembly, figure 6-5c, the liner, whose inside diameter is eccentric to the rotor, compresses blades 1 and 2 in an inward direction. The volume between the blades is reduced, and pressurized fuel is delivered through the groove of the regulator assembly, through the rotor, and past the rotor retainers into a channel leading to the hydraulic head passages. Volume between blades continues to decrease, pressurizing the fuel in the quadrant, until blade 2 passes the groove in the regulating assembly.

REGULATOR ASSEMBLY OPERATION

Figure 6-6 shows the operation of the pressure-regulating piston while the pump is running. Fuel output from the discharge side of the transfer pump forces the piston in the regulator against the regulating spring. As flow

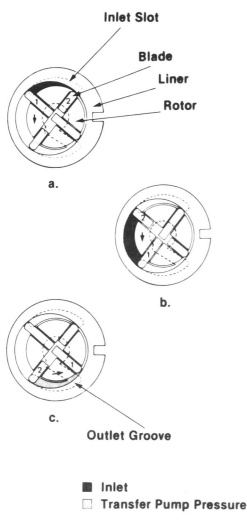

Figure 6-5 Fuel flow through the transfer pump

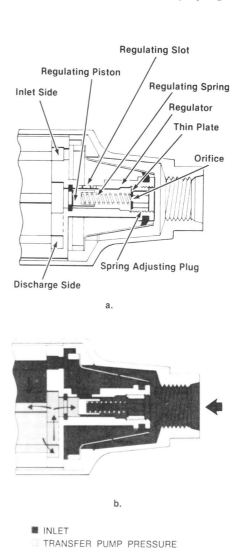

Figure 6-6 Cross section of regulator assembly and fuel flow. (*Courtesy of Stanadyne Diesel System*)

increases, the regulating spring is compressed until the edge of the regulating piston begins to uncover the pressure-regulating slot, figure 6-6b. Since fuel pressure on the piston is opposed by the regulating spring, the delivery pressure of the transfer pump is controlled by the spring rate and size of regulation slot "A." Therefore, pressure increases with speed. A high-pressure relief slot "B" is incorporated in some regulators as part of the pressure-regulating slot to prevent excessively high transfer pump pressure in case the pump or engine is accidentally overspeeded.

VISCOSITY COMPENSATION

Different grades of diesel fuel and varying temperatures, both of which affect fuel viscosity, are compensated for to provide smooth engine performance. A unique and simple feature of the regulating system offsets pressure changes caused by viscosity differences. Located in the spring-adjusting plug is a thin plate incorporating a sharp-edge orifice. The orifice allows fuel leakage past the piston to return to the inlet side of the pump. Flow through a short orifice is virtually unaffected by viscosity changes. The biasing pressure exerted against the back side of the piston is determined by the leakage through the clearance between the piston and regulator bore and the pressure drop through the sharp-edged orifice. With cold or viscous fuels, very little leakage occurs past the piston. The additional force on the back side of the piston from the viscous fuel pressure is slight and the piston will be pushed

Figure 6-8 The piston has covered part of the regulating slot.

further back into the regulating bore, exposing more of the regulating slot, figure 6-7. With hot or light fuels, leakage past the piston increases. Fuel pressure in the spring cavity increases also, since flow past the piston must equal flow through the orifice. Pressure rises because of increased fuel leakage past the piston and the limited amount of fuel that can flow through the orifice. The increased fuel pressure behind the piston assists the regulating spring in keeping the piston from moving back into the bore, exposing little of the regulating slot, figure 6-8. Variations in piston position compensate for changes in the fuel's viscosity, allowing the pump to maintain specified pressures.

CHARGING AND DISCHARGING

Charging Cycle

As the rotor revolves, figure 6-9, the two inlet passages in the rotor register with parts of the circular charging passage. Fuel under pressure from the transfer pump, controlled by the opening of the metering valve, flows into the pumping chamber forcing the plungers apart. The plungers move outward, a distance proportionate to the amount of fuel required for injection on the following stroke. If only a small quantity of fuel is admitted into the pumping chamber, as at idling, the plungers move out a short distance. Maximum plunger travel, and consequently maximum fuel delivery, are limited by the leaf spring that contacts the edge of the roller shoes. Only when the engine is operating at full load will the plungers move to the most outward position. Note in figure 6-9

Figure 6-7 Regulating slot is fully exposed, bypassing fuel back to inlet side.

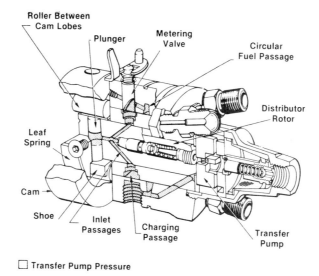

☐ Transfer Pump Pressure

Figure 6-9 Cross section of hydraulic head and rotor during the charging cycle. *(Courtesy of Stanadyne Diesel System)*

that while the angled inlet passages in the rotor are in registry with the ports in the circular charging passsage, the rotor discharge port is not in registry with a head outlet. Note also that the rollers are off the cam lobes.

Discharging Cycle

As the rotor continues to revolve, figure 6-10, the inlet passages move out of registry with the charging ports. The rotor discharge port opens to one of the head outlets. The rollers then contact the cam lobes, forcing the shoes in against the plungers, and high-pressure pumping begins.

Beginning of injection varies according to load (volume of charging fuel), even though rollers may always strike the cam at the same position. Further rotation of the rotor moves the rollers up the cam lobe ramps, pushing the plungers inward. During the discharge stroke, the fuel trapped between the plungers flows through the axial passage of the rotor and discharge port to the injection line. Delivery to the injection line continues until the rollers pass the innermost point of the cam lobes and begin to move outward. The pressure in the axial passage is then reduced, allowing the nozzle to close. This is the end of delivery.

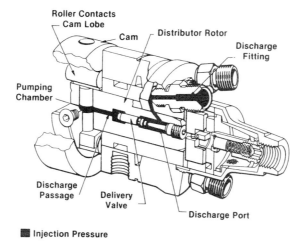

■ Injection Pressure

Figure 6-10 Discharge cycle. High-pressure fuel has forced the delivery valve off its seat. *(Courtesy of Stanadyne Diesel System)*

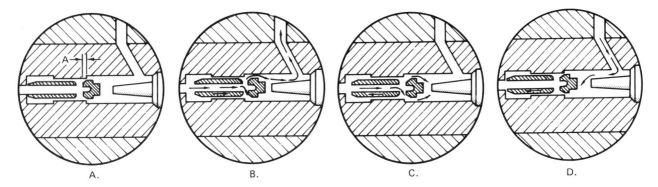

Figure 6-11 Delivery valve operation *(Courtesy of Stanadyne Diesel System)*

DELIVERY VALVE

The delivery valve, figure 6-11a–d, rapidly decreases injection line pressure after injection to a predetermined value lower than that of the nozzle-opening pressure. This reduction in pressure permits the nozzle valve in the nozzle to return rapidly to its seat, achieving sharp delivery cut-off and preventing improperly atomized fuel from entering the combustion chamber.

The delivery valve operates in a bore in the center of the rotor. Note that the valve does not require a seat, only a stop to limit travel. Sealing is accomplished by the close clearance between the valve and the bore into which it fits. Since the delivery valve performs the function of retraction for each injection line, the result is a smooth-running engine at all loads and speeds.

When injection starts, fuel pressure moves the delivery valve slightly out of its bore and adds the volume of its displacement, figure 6-11a, to the delivery valve spring chamber. Since the discharge port is already opened to a head outlet, the retraction volume and plunger displacement volume are delivered under high pressure to the nozzle. Delivery ends when pressure on the plunger side of the delivery valve is quickly reduced, because of the cam rollers passing the highest points on the cam lobes.

Following this, the rotor discharge port closes completely, and a residual injection line pressure is maintained. Note that the delivery valve is only required to seal while the discharge port is open. Once the discharge port is closed, residual injection line pressures are maintained by the seal of the extremely close fit between the head and rotor.

Trailing Port Snubber Operation

PINTLE TYPE NOZZLES

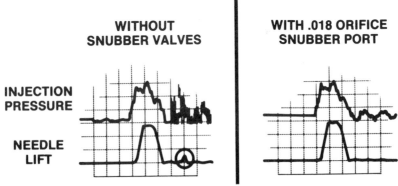

Figure 6-12 Pressure waves are dampened with the use of a snubber valve preventing secondary injections. *(Courtesy of General Motors Product Service Training)*

Snubber Valve and Trailing Port Snubber

A snubber valve and trailing port snubber (TPS) may be used on some models to prevent secondary injections and cavitation erosion of the high-pressure system. Secondary injection is the unwanted injection of fuel into the combustion chamber after pump delivery has ended. When pump delivery ends and the delivery valve suddenly closes, a high-pressure wave is reflected back to the injection nozzle and has sufficient force to push the nozzle valve off its seat, allowing unwanted fuel to enter the combustion chamber. The snubber valve weakens the reflected pressure waves preventing secondary injection, figure 6-12. A flat check, orifice-type snubber valve is installed in each discharge fitting. During the pumping event, the snubber valve opens and allows unrestricted flow to the injection line. At the end of pumping, the snubber valve is closed by an upstream pressure drop (toward the pump). An orifice in the snubber valve provides a restriction that reduces the force of the pressure wave and prevents

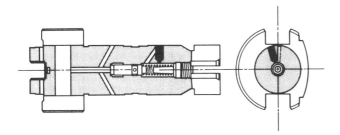

Figure 6-13 Rotor with trailing port snubber *(Courtesy of General Motors Product Service Training)*

RETURN OIL CIRCUIT

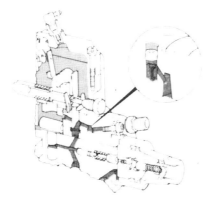

Figure 6-14 Location of vent wire assembly. *(Courtesy of General Motors Product Service Training)*

cavitation erosion. The other type of snubber valve is called the trailing port snubber (TPS) valve. This port trails the delivery port, figure 6-13, and partially deflects the pressure waves through an orifice .018 in. (.457 mm) to the delivery valve cavity. There is only one TPS directly behind the discharge port in the rotor. The TPS reduces pressure, one line at a time, directly after injection.

Those rotors without TPS use residual balancing ports (vented rotor) in which there are as many ports as there are cylinders. These ports align simultaneously after each injection, balancing the residual pressure in all lines.

RETURN FUEL CIRCUIT

Fuel under transfer pump pressure is discharged into a vent passage in the hydraulic head, figure 6-14. Flow through the passage is restricted by a vent wire assembly to prevent excessive return oil and undue pressure loss. The amount of return oil is controlled by the size of the wire used in the vent wire assembly, that is, the smaller the wire the greater the flow, and vice versa. The vent wire assembly is available in several sizes to meet the specified return oil quantities. Note that this assembly is accessible by removing the governor cover. The vent passage is located behind the metering valve bore that connects with a short vertical passage containing the vent wire assembly and leads to the governor compartment.

Should a small quantity of air enter the transfer pump, it immediately passes to the vent passage as shown. Air and a small quantity of fuel then flow from the housing to the fuel tank via the return line.

Housing pressure is maintained by a spring-loaded ball-check return fitting in the governor cover of the pump.

MIN-MAX GOVERNOR

The governor mechanism consists of a cage with flyweights mounted on the rotor, and a system of linkages that control engine speed at idle and rapid governor cut-off to prevent engine overspeed. At all other speeds and loads, however, it acts as a solid link between the accelerator pedal and the metering valve. To accomplish this, two governor springs are used. The smaller, weaker spring compresses at engine idle speed (pump speed is one-half engine speed). Deflection of the larger main governor spring occurs at *engine cut-off speed*. Hence it is called "min-max," indicating the governor controls fuel delivery at minimum and maximum speeds only.

The low throttle position view, figure 6-15a shows the relationship of the governor parts when the pump is running at idle. With the throttle shaft in low-idle position, the balance between the idle spring force and the governor weight force positions the metering valve for low-idle fuel delivery.

In the middle range, the large spring is not compressed. In this position, the driver directly controls the metering valve, figure 6-15b.

In the full throttle position view shown in figure 6-15c, the governor is in the high-speed cut-off position. With the throttle in full position and the engine without a load, pump speed increases until the governor weights have generated enough force to deflect the main governor spring. Governor arm movement turns the metering valve to the shut-off position, restricting fuel delivery and preventing engine overspeed.

FUEL TEMPERATURE COMPENSATOR FOR IDLE SPEED

When the injection pump is located in the center of a V-type engine, heat retention is a critical factor. The high temperatures lower the fuel's viscosity. As the hotter, thinner fuel passes through the pump, internal leakage increases, reducing pump output. To prevent the engine from stalling from too small an output, a bimetal compensator strip, figure 6-16, is added to the governor arm. As temperatures increase, the bimetal compensator strip deflects relative to the governor arm. This motion increases the metering valve opening and provides a compensated idle speed curve.

AUTOMATIC ADVANCE MECHANISM

The automatic advance is a hydraulic mechanism that is powered by fuel pressure from the transfer pump to rotate the cam slightly and vary delivery timing. The advance mechanism advances or retards start of fuel delivery in response to engine speed changes. Without an advance mechanism, the actual beginning of delivery of fuel at the injector starts later (in engine degrees of retation) as speed increases.

Compensating inherent injection lag improves high-speed performance of the engine. Starting delivery of fuel to the injector earlier, when the engine is operating at a higher speed, ensures that combustion takes place when the piston is in its most effective position to produce optimum power with a minimum of fuel and smoke.

The advance pistons located in a bore at the bottom of the housing engage the cam advance screw and move the cam (when fuel pressure moves the power piston) opposite the direction of rotor rotation, figure 6-17a–b. Fuel under transfer pump pressure is fed through a drilled

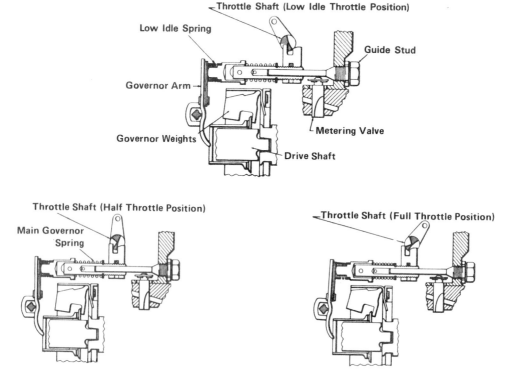

Figure 6-15 Min-max governor operation (*Courtesy of General Motors Product Service Training*)

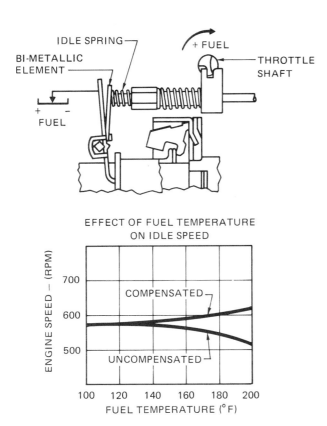

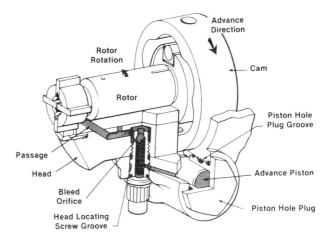

Figure 6-16 Bimetallic element compensates for the increase in fuel temperature by putting more pressure on the idle spring thus increasing engine idle. (*Courtesy of General Motors Product Service Training*)

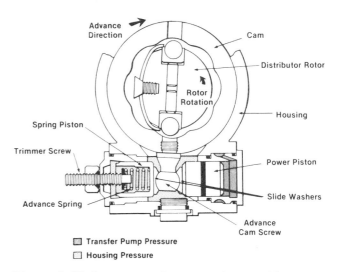

Figure 6-17 Automatic advance mechanism (*Courtesy of Stanadyne Diesel System*)

passage in the hydraulic head that registers with the bore of the head-locating screw. Fuel is then directed past the spring-loaded ballcheck in the bore of the head-locating screw. It then enters the groove on the outside diameter of the screw, which registers with a drilled passage in the housing leading to the power side of the automatic advance assembly.

A groove around the power piston plug and a drilled passage allow the fuel to enter the advance piston bore. Fuel pressure against the piston must overcome the opposing spring force, plus the dynamic injection loading on the cam, in order to change the cam position. The spring-loaded ballcheck in the bore of the head-locating screw prevents the normal tendency of the cam to return to the retard position during injection by trapping the fuel in the piston chamber. When engine speed decreases, the hydraulic pressure is reduced and the spring returns the cam to a retarded position in proportion to the reduction in speed. The fuel in the piston chamber is allowed to bleed off through a control orifice located below the ballcheck valve in the head-locating screw.

At low speeds, because transfer pump pressure is comparatively low, the cam remains in the retarded position.

When engine speed increases, transfer pump pressure rises and moves the piston in the advance direction. Advance piston movement is related to speed. Total movement of the cam is limited by piston length, figure 6-18.

A trimmer screw is used on some models to adjust spring preload, which controls the start on cam movement. The trimmer screw cannot be adjusted in the field; a special test stand is needed.

CAUTION: Do not touch it.

On many later models, beginning with 1980, in addition to the normal automatic advance, a mechanical/light load advance mechanism was added.

Two subsystems are combined to form the mechanical/light load advance. The first is a servo advance mechanism that is operated by transfer pump pressure and positions the cam ring in response to throttle setting and engine load. The major component parts of the advance are the

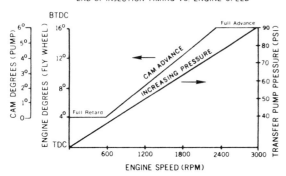

SPEED ADVANCE OPERATION

END OF INJECTION TIMING VS. ENGINE SPEED

Figure 6-18 Injection timing advances as transfer pump pressure and engine RPMs increase. *(Courtesy of Stanadyne Diesel System)*

servo advance piston, the cam advance pin, the servo advance valve, the servo advance plunger, and the mechanical/light load advance spring. This system is housed in the standard DB2 advance boss location and receives transfer pump pressure from the head-locating screw.

The second subsystem is the mechanical link between the throttle shaft and the servo advance plunger. This link is composed of a face cam connected to the end of the throttle shaft and a rocker lever assembly connected to the side of the pump housing by a pivot pin. A roller is attached to the upper end of the lever and rides on the surface of the face cam. The lower end of the lever contacts the protruding end of the servo advance plunger.

During pump operation, a rotary force is imparted to the cam ring by the cam rollers during injection. It acts in the direction of rotor movement and is transmitted through the cam advance pin to the servo advance piston. This force continually urges the piston toward the retard position.

An opposing force is supplied by transfer pump pressure acting on one end of the servo advance piston. The position of the servo advance valve in the piston bore regulates this force and determines the degree of advance achieved at any throttle setting or load. It is, in turn, the differential between mechanical/light load spring force and transfer pump pressure applied across the servo advance valve that locates the valve in the piston bore. Additional advance at low throttle settings is provided by the face-cam-to-rocker-lever action, which changes the reference point of the spring.

Housing pressure acting on the spring-side end of the advance piston forms a resistance to its movement at all speeds. If housing pressure is reduced at speeds below full advance, further advance piston movement occurs.

To help understand the description of mechanical/light load advance operation, refer to figure 6-19a–f.

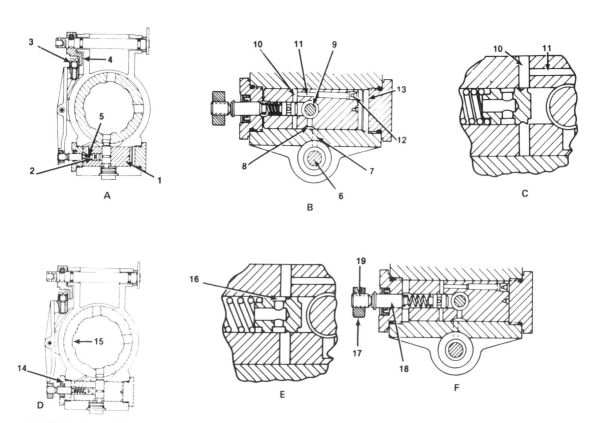

Figure 6-19 Mechanical/light load advance mechanism *(Courtesy of General Motors Product Service Training)*

Figure 19a shows the cam ring and advance mechanism in the retard position as it would be at cranking. The advance piston (1) is moved toward the power plug, and the servo advance valve (2) is in the position shown. With the throttle closed, the roller (3) rides on the low portion of the face cam (4) and no change is made in the reference point of the mechanical/light load advance spring (5).

As the engine accelerates, increasing transfer pump pressure is directed through the head-locating screw (6) to the housing passage (7), shown in 19b. This passage empties into an elongated groove (8) in the advance piston. Fuel then flows around the groove to the cam advance pin (9) and into the piston bore.

With the servo valve in the position shown in figure 19c, fuel enters the transverse passages (10) in the advance piston and fills the single longitudinal passage (11) that extends to the piston end. An orifice screw (12) is located at this point to restrict the flow of fuel and eliminate fluctuations caused by the varying amounts of cam loading during each pumping cycle. Fuel then fills the cavity between the advance piston and the power piston plug (13); the pressure acts on the surface of the piston and urges it, against cam loading force, toward the spring-side plug (14), shown in 19d. This linear piston motion is converted to a rotary motion of cam (15), which advances the beginning of pumping. Advance piston movement and cam ring rotation continues as long as pump speed and transfer pump pressure increase. When pump speed stabilizes, the force of cam loading balances the power piston force, and piston movement ceases. The servo valve, which has been moved toward the spring-side end plug by the force of transfer pump pressure, tends to hover over the transverse passage because of the balancing spring force. At stable pump speed, the valve's forward edge contacts the edge of the transverse passage, sealing transfer pump pressure in the longitudinal passage and piston cavity. Increased pump speed, or transfer pump pressure, again moves the valve against spring force to open the transverse passage and permit further advance action.

If pump speed decreases, transfer pump pressure also decreases. The reduced force on the servo valve allows the advance spring to move the valve toward the cam advance pin and align the transverse passages (16) in the valve (see 19e) with the passages in the piston. Fuel then flows through the valve and into the piston bore. Pressure in the power-plug side piston cavity is reduced, and the cam loading force urges the valve toward the retard position as shown in 19f. Fuel in the piston bore is vented into the housing. This action continues until the movement of the piston, relative to the now stationary servo valve,

repositions the two parts, as shown in 19c, and the piston transverse passages are opened to transfer pump pressure.

During the initial degrees of throttle travel, the face cam and lever assembly (17) positions the advance spring at its outboard location. This permits greater movement of the advance piston and cam ring before the servo valve begins to close the transverse passages; increased advance action at low throttle settings is then achieved. As the throttle travel angle is expanded, the ramp on the face cam actuates the lever assembly, moving the plunger (18) and the spring reference point toward the cam pin.

At a predetermined throttle angle, the surface of the face cam becomes a flat plane and the spring reference point becomes fixed. Advance action after this point is regulated by the spring rate. It should be emphasized that the action of the face cam and lever assembly does not change the rate of the action spring or its loading, only its reference point.

A system-adjusting screw (19) is provided in the plunger end of the lever assembly.

HOUSING PRESSURE COLD ADVANCE (HPCA)

The HPCA feature is designed to advance the injection timing about 3 degrees during cold operation. Its main purpose is the reduction of engine smoke, roughness, noise, and emissions during cold start-up by advancing the fuel delivery system. Through the use of an engine-mounted temperature switch (the same switch that operates the fast idle solenoid below 125°F [52°C]), a solenoid located in the pump cover pushes the return fitting checkball away from its seat (Refer to figures 6-1 and 6-2). This relieves the normally 10-psi (68.95-kPa) housing pressure, which in turn relieves some of the pressure of opposing the timing advance piston. Transfer pump pressure advances the timing an additional 3 degrees, which initiates combustion sooner and results in a slower, more complete burning of the fuel. Above 125°F (52°C), the switch opens, de-energizing the solenoid; the checkball returns to its seat, and housing pressure is returned back to 10 psi (68.95 kPa). The switch again closes when temperature falls below 95°F (35°C). This cold idle and starting device, depending on engine temperature, operates from 2 to 10 minutes.

ELECTRIC SHUT-OFF

The pump is equipped with an electric shut-off solenoid. When no voltage is present, the solenoid, which is connected to the governor linkage, pushes the metering valve into the closed position, preventing fuel flow. When the solenoid is energized, spring pressure on the solenoid arm is overcome. Through the governor linkage, the metering valve operates freely throughout its throttle range.

Coil temperature has an effect on the pull-in voltage required to operate the solenoid. Figure 6-20 shows the maximum production limits. This pull-in voltage requirement should always be considered when diagnosing no-start conditions. Low battery voltage may be the cause. Pull-in voltage can be tested with a voltmeter at the pump solenoid during normal starting operation. Pull-in voltage should be approximately 8.8 volts maximum.

SERVICING THE ROOSA MASTER DB2 FUEL INJECTION PUMP

Working on fuel injection components requires the utmost care and cleanliness. All tools, equipment, work areas, and parts must be kept extremely clean, as must technicians' hands. Only lint-free towels should be used.

All fasteners must be torqued to manufacturer's specifications, NO EXCEPTIONS.

CAUTION: This model pump is used by different manufacturers to meet their needs.

Each manufacturer has a limit on how much the pump can be serviced. Consult the manufacturer's shop manual. Failure to do so may void any warranty on the pump.

You will also need the special tools required to perform work on this pump. Again, consult the manufacturer's service manual.

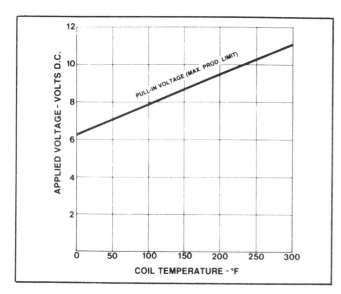

Figure 6-20 Required pull-in voltage increases as coil temperature increases. *(Courtesy of General Motors Product Service Training)*

CAUTION: The following procedures are intended to be used as a guide only, and meant to explain what you will be doing. Whenever performing any service work on the pump, use the detailed service procedures given in the manufacturer's service manual.

Pump Removal and Installation

Removal. This task requires the following:

- appropriate hand tools
- manufacturer's special tools
- manufacturer's service manual

To remove the injection pump, certain components must be removed. Usually, the air cleaner assembly, and if present, the air crossover assembly are removed, exposing unrestricted openings to the intake manifold.

CAUTION: Always cover the air intakes to prevent debris from entering the cylinder.

Throttle linkages and hoses must be disconnected. Note where and how they were removed.

Disconnect injection lines at the injectors, inlet fuel line, return fuel line, and cap all openings.

Mark the position of the pump to the engine. This will aid in keeping timing correct when installing the pump.

Disconnect the pump and carefully lift it out of the engine.

CAUTION: Do not bend or kink any fuel lines.

Discard old O ring.

CAUTION: Do not crank engine unless the manufacturer states that you can.

The reason is that some timing drives are gear driven. Rotating the engine with the pump out could cause the gears to jump out of time. When shipping the pump for repairs, be sure all exposed fuel openings are capped.

Installation. The tang on the end of the pump drive shaft is offset, allowing the pump to fit one way, Figure 6-21. Therefore, it is necessary to position either the pump drive shaft, the pump drive, or both, for them to line up properly, figure 6-22. If the engine cannot be rotated, put the pump drive shaft in the same position as when the pump was removed.

Figure 6-21 Offset tang on pump driveshaft

Always use a new pump adapter O ring. Install the pump by hand. Be sure it is fully seated. Do not attempt to tighten the pump if it is not seated.

Line up the pump timing marks and tighted the pump to specifications.

Connect the fuel and injection lines and tighten to specifications. Undertightening can cause leaks and poor engine performance.

Install the throttle cables and return springs.

CAUTION: Always check the throttle assembly to be certain it does not bind or stick.

Start the engine and check for leaks. Check idle speed and pump timing. If out of specification, follow the adjustment procedures.

Remove the protective screens, install the air crossover and air cleaner.

Injection Pump Idle Speed Adjustments

This task requires the following:

- appropriate hand tools.
- manufacturer's special tools (magnetic pick-up meter to measure engine RPM)
- manufacturer's servive manual

The DB2 pump is equipped with a slow-idle adjustment screw.

CAUTION: This is the only screw that is adjusted in the field. All other adjustments on the pump itself are handled at authorized dealers.

For this procedure, the vehicle must be prepared according to the manufacturer's directions. For example, all electrical accessories must be turned off and the vehicle fully warmed up.

To measure engine RPM, a special tachometer is needed. A magnetic probe is inserted in a probe holder just above the crankshaft balancer. Cut into the balancer is a narrow groove. This groove is sensed by the magnetic probe each time it passses under the probe. A digital readout indicates how many times the groove is passing under the probe, which is engine RPM. Follow the manufacturer's directions when installing the tachometer. Clean

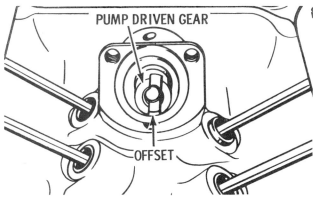

Figure 6-22 Offset slot on pump-driven gear (*Courtesy of Oldsmobile Division*)

the probe holder and crankshaft balancer rim to prevent false readings.

With the car properly prepared and the tachometer hooked up, check slow-idle speed reading against the one given on the VEHICLE EMISSION INFORMATION LABEL. Reset if required, figure 6-23.

To help the engine run when it is cold, a fast-idle circuit is used. An engine temperature (fast-idle cold advance) switch turns on at a preset level (approximately 125°F [52°C]) and sends current to a fast-idle solenoid that increases engine speed. With the engine warmed up it is necessary to bypass the fast-idle cold advance (engine temperature) switch and install a jumper between the connector terminals. Do not allow the jumper wire to touch the ground, figure 6-24.

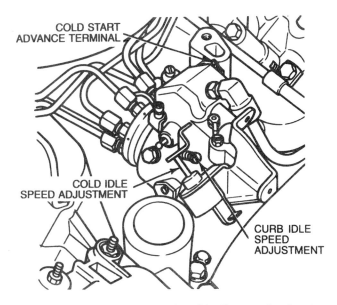

Figure 6-23 Curb idle and cold idle speed adjusting screws (*Courtesy of Ford Motor Company*)

To achieve a fast idle when the engine is cold, an electric solenoid, mounted on the injection pump, is connected to an engine temperature switch.

Check the fast-idle solenoid speed against the one given on the VEHICLE EMISSION INFORMATION LABEL. Adjust if required, figure 6-25.

Remove the jumper lead and connect the temperature switch. Check again and reset the slow-idle speed, if necessary.

Install all components that were removed and be sure the vehicle is ready for the owner.

Checking and Adjusting Timing

Checking and adjusting injection pump timing on a diesel engine requires a different timing meter and procedures that may be unfamiliar to you. This task requires the following:

- appropriate hand tools
- special timing meter
- manufacturer's service manual

Figure 6-25 Fast-idle solenoid (*Courtesy of Oldsmobile Division*)

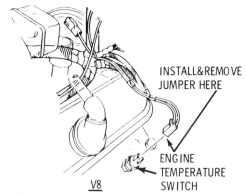

Figure 6-24 Jumper wire location to set fast idle (*Courtesy of Oldsmobile Division*)

To measure timing on a gasoline engine, a timing light picks up the electrical pulse of the NO. 1 spark plug wire, then flashes a strobe light on a set of timing marks. The numbered mark that lines up with the pointer indicates how many degrees, before or after TDC, the spark reached the cylinder. Timing is adjusted by simply rotating the distributor. Of course, with a diesel, there is no electrical spark to sense, so alternate methods must be used. Two forms of energy that are sensed are (1) the light of combustion, and (2) the injection line pulse.

The light-of-combustion method uses a luminosity probe that is inserted into the designated glow plug hole. A light-sensitive photo detector sees the light of combustion and sends a signal to the timing meter. The timing meter also senses engine RPM through a magnetic probe inserted in the probe holder. The timing meter processes the light-of-combustion event and engine RPM simultaneously, to give a readout of how many degrees the injection pump is advanced or retarded, figure 6-26.

When the luminosity probe is inserted into the glow plug cavity while the engine is running, the light of combustion is visible.

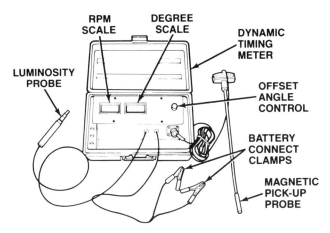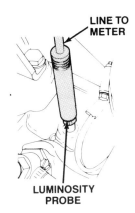

Figure 6-26 Luminosity-type meter for determining injection pump timing (*Courtesy of Ford Motor Company*)

The injection line pulse method uses a transducer that is clamped over the designated injector line. The transducer senses the high-pressure fuel pulse and converts this into an electrical signal. The timing meter processes this signal as well as engine RPM from the magnetic probe to give a timing reading.

Regardless of which type you use, follow the manufacturer's directions explicitly. Any deviation may result in false readings. The following are important tips to prevent false readings:

- Always clean the magnetic probe holder and crankshaft balancer. Do not score or scratch the balancer rim.
- Be sure to use the proper glow plug hole or injection line. It is not always no. 1 cylinder.
- Use the proper offset angle.
- Be sure the luminosity probe is clean and free of soot. Use a toothpick to clean the luminosity probe and a cotton swab to clean the photo detector lens.
- Prepare the vehicle according to the manufacturer's directions.
- Wait for the readings to stabilize and check at 2-minute intervals.
- Rapid needle fluctuations on the timing meter indicate that the cylinder is misfiring.

CAUTION: Never attempt to adjust the pump with the engine running. Always shut the engine off.

Moving the pump while the engine is running can cause the pump to seize. The clearance between the hydraulic head and rotor is so close that if the pump were moved to one side, the rotor would contact the head and seize.

When you have taken a timing reading, shut the engine off, loosen the pump, and rotate it either right or left. To rotate the pump, insert an open-end wrench on the pump boss, which is located at the front of the pump. Rotate the pump the amount needed and tighten the pump to specifications. Start the engine and check the timing.

Vacuum Regulator Valve

The diesel engine produces very little intake manifold vacuum. Unfortunately for the diesel, automatic transmissions and emission control devices work off intake manifold vacuum. To compensate for this, a vacuum pump is added to produce the necessary vacuum. The vacuum regulator valve (VRV), which is mounted on the side of the injection pump, regulates vacuum according to throttle position, figure 6-27. Together they imitate the vacuum readings that would be found on a gasoline engine.

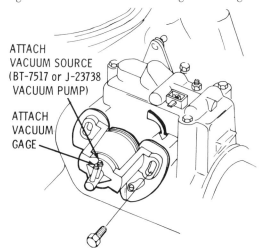

Figure 6-27 Vacuum regulator valve (*Courtesy of Oldsmobile Division*)

Hooked into the VRV is the throttle shaft. At idle, vacuum output is highest from the VRV, and vacuum steadily decreases to zero at the full fuel position. By controlling the amount of vacuum output according to throttle position, the VRV is useful in sending the proper vacuum signal to the transmission and emission control devices.

The VRV is adjustable and replaceable. This task requires the following:

- appropriate hand tools
- manufacturer's special tools (carburetor angle gauge and adapter)
- vacuum pump
- manufacturer's service manual

First, do not remove or move the VRV unless unavoidable. Second, if the VRV must be taken off and put back on the same injection pump, it should be marked so that it is installed exactly the way it came off.

The VRV adjustment is needed when a new injection pump is being installed, and when the VRV has been moved. It may also be performed if the customer complains of poor transmission shift quality.

Follow the manufacturer's guide for installing the adapter and carburetor angle gauge, figure 6-28. With the angle gauge in the proper position, a vacuum pump is hooked into the inlet of the VRV and also at the output side. The VRV is rotated until the vacuum gauge registers the specified reading. Precision is needed on this adjustment because the amount of vacuum output is being set at a specific throttle position.

Injection Pump Housing Pressure Check

The injection pump housing pressure check is made if the idle is rough or the engine starts but will not run. This task requires the following:

- appropriate hand tools
- manufacturer's special tools (pressure tap adapter and pressure gauge)
- tachometer
- manufacturer's service manual

Follow the manufacturer's directions for removing the pressure tap plug or torque screw.

CAUTION: Do not disturb torque screw adjustment.

If it is disturbed, the pump must be sent to an authorized shop for calibration.

Install the pressure tap adapter, pressure gauge, and tachometer, figure 6-29. With the engine running and the drive wheels blocked, set the engine to the specified speed and note the pressure gauge reading (usual specifications are at 1000 RPM gauge and should read 8–12 psi [55.16–82.74 kPa] with no more than 2 psi [13.79 kPa] fluctuation).

If pressure is zero, the HPCA must be checked.

If pressure is low, replace the return fuel line connector.

If pressure is too high, there is either a restriction in the fuel line return system, the return fuel line connector is bad, or the pump must be sent out for repair. First, disconnect the return fuel line and run a hose into a

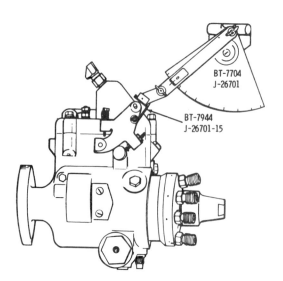

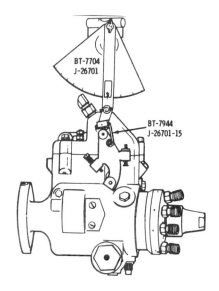

Figure 6-28 Adjustment of VRV using angle gauge *(Courtesy of Oldsmobile Division)*

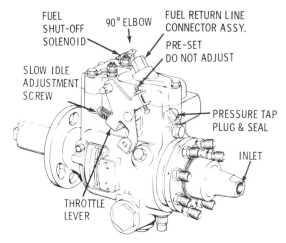

Figure 6-29 External features of injection pump with location of pressure tap plug and seal (*Courtesy of Oldsmobile Division*)

container. If pressure returns to normal, find and repair the restriction.

If housing pressure is still too high, replace the return fuel line connector.

If housing pressure remains high, send the pump out for repair.

Governor Weight Retainer Ring

The governor weight retainer ring, located in the governor weight retainer, is made of a special plastic (Pellethane) designed to absorb shock and vibration between the drive shaft and governor weight retainer. Rough idle or failure to start may be symptoms of this problem when the ring fails. Ring failure results from the high temperatures in the cavity the injection pump occupies. Ring failure can also be accelerated by the use of alcohol and other additives.

Inspecting for this condition requires the following:

● appropriate hand tools
● manufacturer's special tools
● manufacturer's service manual

With the air cleaner assembly and crossover assembly removed, install screens to cover exposed openings.

Disconnect the fuel solenoid wire, HPCA wire, and fuel return line.

CAUTION: Clean the pump cover and surrounding area to prevent dirt from entering the pump. Place several rags under the pump to catch excess fuel.

Remove the injection pump cover.

CAUTION: Do not alllow any dirt to enter the pump with the cover off. Damage will occur if foreign material is left in the pump.

With the pump cover removed, you will be able to see the governor weight retainer. With a screwdriver, try to rotate the weight retainer in both directions, figure 6-30. If the retainer moves more than 1/16 in. (1.65 mm) and does not return, the ring has failed. The injection pump must be removed and disassembled or sent to an authorized service dealer for repair. When installing the pump cover, install it at an angle to prevent the linkage from binding, figure 6-31.

Injection Pump Disassembly

This task requires the following:

● appropriate and clean hand tools
● manufacturer's special tools (injection pump service kit)
● special mounting fixture
● seal kit
● manufacturer's service manual

No directions on how to disassemble and assemble the pump will be given here because the directions are lengthy and MUST BE STRICTLY FOLLOWED TO ENSURE THAT THE PUMP WILL WORK. The following tips can help ensure success.

Be sure the work area and tools are extremely clean. Remember that dirt particles larger than 10 microns will damage the pump.

Use lint-free towels. Even lint will damage the pump.

CAUTION: Do not disturb any of the adjustments. Any adjustment on your part will void any obligation and warranty by the manufacturer.

If any adjustment has been disturbed, the pump must be sent out for calibration.

Handle only those surfaces designated in the manual. The precision parts in the pump are sensitive to corrosion.

Torque all bolts to the specified values with an accurate torque wrench.

Check throttle shaft and metering valve movement.

Test the pump for leaks when assembly has been completed, and check housing pressure when the pump has been installed on the vehicle.

Injection Pump Leak Testing

This task requires the following:

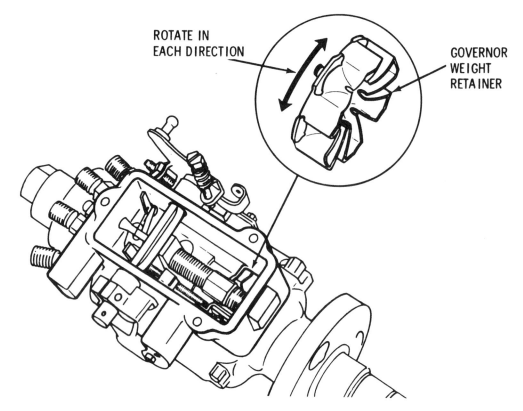

Figure 6-30 Location of governor weight retainer *(Courtesy of General Motors Product Service Training)*

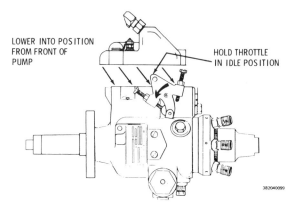

Figure 6-31 The correct way to install the pump cover *(Courtesy of Oldsmobile Division)*

- appropriate hand tools
- regulated air pressure
- container of clean fuel, clean calibrating fluid, or light mineral seal oil
- manufacturer's service manual

Test the injection pump for leaks by following these procedures:

1. Be sure that the pump is well drained of fluid.
2. Plug the return line connector assembly.
3. Install the appropriate fitting in the transfer pump inlet to allow the connection of a clean, filtered air supply.
4. Connect the air supply at 10–15 psi (68.95–103.42 kPa) maximum to the fitting installed in the pump inlet.
5. Completely submerge the pump in clean fluid.
6. Allow the pump to remain submerged for a minimum of 5 minutes. Check for leaks. Note: Some air may leak from the fuel line connectors. This is normal and not cause for repair. Rotating the drive shaft slightly to position the head and rotor parts out of registry will reduce air leakage.
7. Repair leaks as necessary and repeat the leak test.

CAUTION: On pump models where the drive shaft is retained by an "O" ring, it is possible that the shaft will pop out during the leak test procedure. To prevent this, a device similar to the one shown in figure 6-32A may be fabricated from 3/32-inch (.093) wire and installed as show in figure 6-32B.

INJECTION PUMP CALIBRATION

When the injection pump is sent to an authorized Roosa Master dealer, it is first inspected visually for physical damage. If repair is indicated at this time, it is done

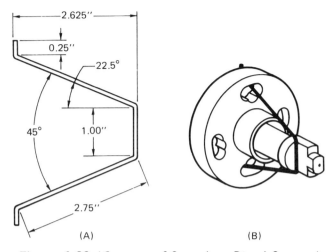

Figure 6-32 *(Courtesy of Stanadyne Diesel Systems)*

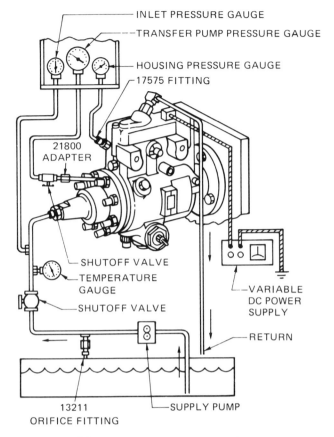

Figure 6-33 Injection pump mounted on test stand
(Courtesy of General Motors Product Service Training)

before testing the pump. The pump is then mounted on a fuel test stand, an expensive machine ($20,000–$40,000) that can accurately measure the pump's fuel output under specified conditions.

With the pump mounted on the stand, a coupler is hooked to the pump drive shaft. The pump's discharge outlets are connected each to their own line that leads to their own graduated cylinder. Each graduated cylinder indicates the fuel output from each discharge outlet. Other adapters are hooked to measure transfer pump pressure, housing pressure, and inlet presure, figure 6-33.

Some of the tests performed are: electric solenoid pull-in voltage, HPCA solenoid operation, face cam position, min-max governor operation, return fuel volume, maximum fuel delivery, housing pressure adjustment, transfer pump pressure adjustment, and automatic advance adjustment. All these tests are conducted at specified RPM and throttle positions. No guess work is involved. The pump either meets the specifications or it is repaired.

SUMMARY

The DB2 injection pump is an opposed plunger, inlet-metered, positive displacement, distributor-type pump. A single pumping element pressurizes the fuel for all cylinders. It meets the requirements of a high-speed diesel engine.

The sections dealing with fuel flow, the pumping cycles, governor operation, and timing operation should be studied closely and read as many times as necessary for comprehension. Knowing how they work will aid in diagnosis.

Servicing the DB2 injection pump requires knowledge, care, and the proper tools. Follow the manufacturer's instructions explicitly to avoid damage to the pump and warranty violations.

CHAPTER 6 QUESTIONS

1. List the main components of the DB2 injection pump.
2. Describe and trace the fuel flow with the aid of a diagram.
3. Explain the operation of the transfer pump.
4. Explain the operation of the regulator assembly.
5. Explain how changes in fuel viscosity are compensated.
6. Explain the charging and discharging cycle.

7. Explain the operation of the delivery valve.
8. Describe fuel flow through the return oil circuit.
9. Explain operation of the min-max governor.
10. Explain operation of the automatic advance mechanism.
11. Explain operation of HPCA.
12. Explain operation of electric shutoff.
13. List cautions when servicing the DB2 pump.
14. Describe VRV operation.

chapter 7
The Robert Bosch
VE-Type Distributor Pump

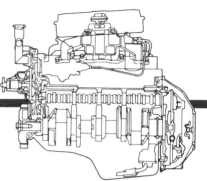

Objectives

In this chapter you will learn:

- To identify the Robert Bosch VE-type injection pump
- To identify its main components
- The pattern of the fuel flow
- The operation of the transfer pump and regulator assembly
- The intake and injection stroke
- The purpose of the equalizing stroke
- How fuel quantity is controlled
- The term *effective stroke*
- The operation of the delivery valve
- The operation of the min-max governor
- The operation of the timer mechanism
- The operation of the cold start device
- The purpose of the aneroid compensator
- How the fuel cut-off solenoid works
- How to perform specified services on the injection pump

IDENTIFYING THE ROBERT BOSCH VE-TYPE INJECTION PUMP

The Robert Bosch VE-type injection pump is used on a variety of vehicles and is made under license by other manufacturers (Diesel Kiki and Nippondenso). It is primarily found on European and Japanese diesel engines. The code designation on the side of the pump is read as follows:

NP–VE x/x F xxxx A R NP xx

a. NP—These two letters stand for the manufacturer, in this case, Diesel Kiki.
b. VE—Distributor-type injection pump.
c. x—number of cylinders
d. x—plunger diameter in millimeters
e. F—mechanical governor
f. xxxx—governor-controlled RPM number
g. A—design symbol
h. R—direction of rotation (R for clockwise, L for counterclockwise)
i. NP xx—production serial number

MAIN COMPONENTS

It is necessary to become familiar with the main components of the VE-type injection pump to understand the basic operating principles. Refer to figures 7-1 and 7-2 for the location of the main components.

A VE pump may be gear or spur belt driven. A pump spur gear is keyed to the pump drive shaft (F), figure 7-2. The drive shaft is connected to a cam plate (K) with a driving disc (I). The cam plate is keyed to the plunger with a knock pin. The drive shaft, cam plate, and plunger all rotate together.

The cam plate has face cams equal to the number of engine cylinders. When the cam plate rotates, it moves side to side in a reciprocating motion, moving the plunger with it. A spring is used to hold the cam plate and plunger against the rollers.

The plunger's reciprocating motion pressurizes the fuel and its rotation distributes the fuel to the proper cylinder.

The drive shaft also drives the fuel supply pump (G) and the governor flyweight drive gear (H). The mechanical governor in the top of the injection pump consists of flyweights (B) contained in a flyweight holder that is mounted on the governor shaft.

The timing device (J) in the bottom of the injection pump is actuated by fuel supply pump pressure. The piston of the timing device moves the roller ring, causing injection timing to be advanced the required amount.

An air bleed screw is fitted in the plug in the middle of the distributor head. Also, a delivery valve (N) for each

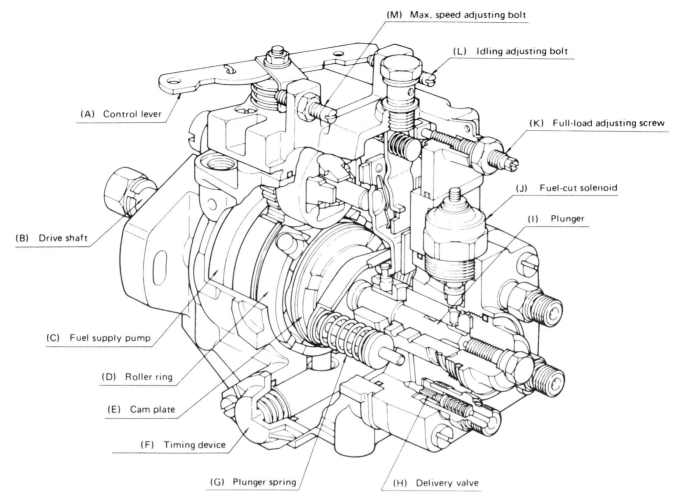

Figure 7-1 Cutaway of VE injection pump showing component location (*Courtesy of General Motors Product Service Training*)

cylinder is mounted in the flange of the distributor head.

A cover is fitted to the top of the injection pump. This cover has a control lever shaft (D) mounted on it. The governor spring (A) is coupled to the control lever shaft (D) via a shackle (V). An idling spring (T) is fitted on the idling pin on the opposite side.

The control lever (C) is fitted to the control lever shaft (D) on the top of the cover. The positions of the control lever (C) are adjusted by means of the maximum speed adjusting bolt and the idle-adjusting bolt to control the respective speeds. Also, a filter and overflow valve (U) are mounted in the top part of the cover.

THE FUEL SUPPLY PUMP

A vane-type fuel supply pump is mounted in the housing on the drive shaft (3) side, figure 7-3. It is used to raise fuel from the tank and feed it under pressure to (a) the distributor plunger, and (b) the injection timing advance mechanism.

When the rotor (2) is driven by the drive shaft (3), the vanes (4) are thrown outward by centrifugal force, forming chambers between adjacent vanes (4). As the rotor (2) turns, the volume of the chamber at the fuel intake side gradually increases, causing the pressure in the chamber to decrease. This allows additional fuel to enter the pump. As the rotor (2) continues to turn, the volume of the chamber starts to decrease, causing the fuel inside to be pressurized. Finally, the fuel is ejected from the outlet under pressure.

THE REGULATING VALVE

The fuel from the fuel supply pump is several times the injection quantity. Excess fuel is returned to the intake side of the fuel supply pump via a regulating valve (B), figure 7-4.

Feed or supply pressure is related to engine speed, and rises as the speed of rotation is increased. A predetermined relationship between feed or supply pressure and pump

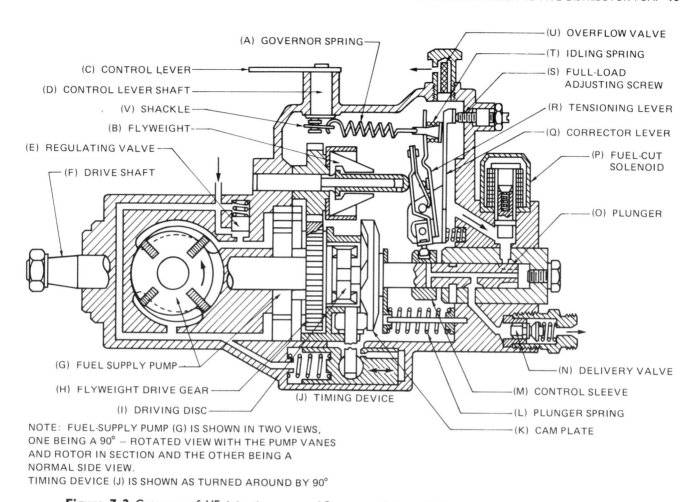

(A) GOVERNOR SPRING
(C) CONTROL LEVER
(D) CONTROL LEVER SHAFT
(V) SHACKLE
(B) FLYWEIGHT
(E) REGULATING VALVE
(F) DRIVE SHAFT
(U) OVERFLOW VALVE
(T) IDLING SPRING
(S) FULL-LOAD ADJUSTING SCREW
(R) TENSIONING LEVER
(Q) CORRECTOR LEVER
(P) FUEL-CUT SOLENOID
(O) PLUNGER
(G) FUEL SUPPLY PUMP
(H) FLYWEIGHT DRIVE GEAR
(I) DRIVING DISC
(J) TIMING DEVICE
(N) DELIVERY VALVE
(M) CONTROL SLEEVE
(L) PLUNGER SPRING
(K) CAM PLATE

NOTE: FUEL-SUPPLY PUMP (G) IS SHOWN IN TWO VIEWS, ONE BEING A 90° – ROTATED VIEW WITH THE PUMP VANES AND ROTOR IN SECTION AND THE OTHER BEING A NORMAL SIDE VIEW.
TIMING DEVICE (J) IS SHOWN AS TURNED AROUND BY 90°

Figure 7-2 Cutaway of VE injection pump (*Courtesy of General Motors Product Service Training*)

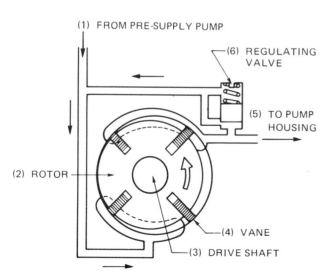

(1) FROM PRE-SUPPLY PUMP
(6) REGULATING VALVE
(5) TO PUMP HOUSING
(2) ROTOR
(4) VANE
(3) DRIVE SHAFT

Figure 7-3 Fuel flow through supply pump (*Courtesy of General Motors Product Service Training*)

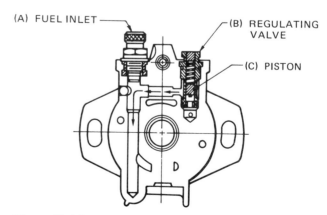

(A) FUEL INLET
(B) REGULATING VALVE
(C) PISTON

Figure 7-4 Regulating valve (*Courtesy of General Motors Product Service Training*)

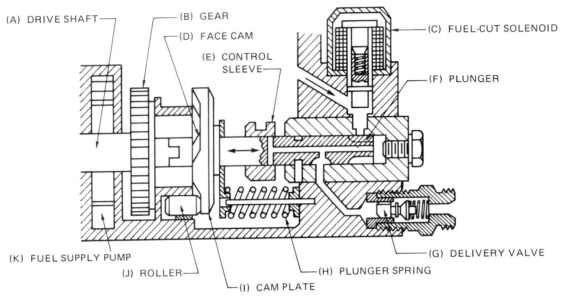

Figure 7-5 Plunger, head, and cam plate assembly *(Courtesy of General Motors Product Service Training)*

speed is maintained by a regulating valve situated in the flange of the pump. The speed/fuel oil pressure characteristics are determined by the spring of the regulating valve (B).

The pressure characteristics will, in turn, control the movement of the piston of the timing device.

PLUNGER OPERATION

The drive shaft (A) simultaneously rotates the fuel supply pump (K), cam plate (I), and plunger (F), figure 7-5. Cam projections are provided on the front surface of the cam plate in the axial direction. When the cam plate rotates, the movement of the cam plate ramp up and down the roller (J) causes the plunger (F) to reciprocate as it rotates. This reciprocating motion pressurizes the fuel and transmits it to the nozzles. Fuel is distributed to each nozzle, as the plunger rotates, figure 7-6.

Details of the complete plunger cycle 1 to 4 are as follows.

Intake Stroke

The intake slit (5) on the plunger (1), figure 7-7, comes into line with the intake port (6) during the intake stroke of the plunger (to the left). The fuel pressurized by the fuel supply pump then flows, into the high-pressure chamber (4) and the plunger body (1).

Injection Stroke

During this stroke, the plunger is pushed to the right as it rotates. First, the intake port closes, causing the

fuel to be compressed, figure 7-8. As the plunger continues moving, the distributing slit (1) on the plunger comes into line with the outlet passage (2), whereupon the pressurized fuel pushes up the delivery valve spring. The fuel is then injected into the combustion chamber of the engine through the nozzle.

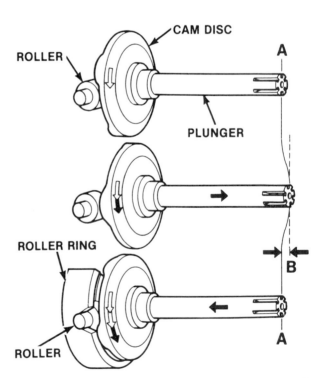

Figure 7-6 It is important to understand that the plunger reciprocates and rotates at the same time. *(Courtesy of Ford Motor Company)*

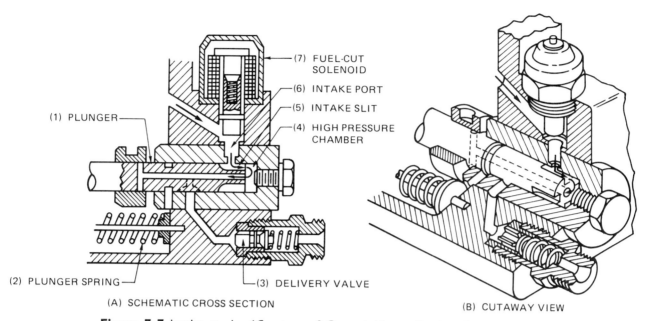

Figure 7-7 Intake stroke *(Courtesy of General Motors Product Service Training)*

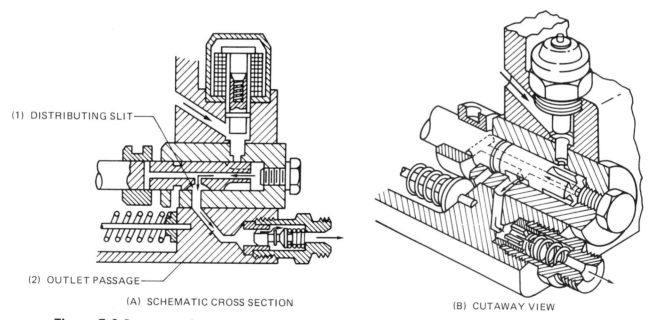

Figure 7-8 Beginning of injection stroke *(Courtesy of General Motors Product Service Training)*

End of Delivery

As the plunger continues moving to the right, figure 7-9, the cut-off port (2) in the plunger is connected with the pump housing, causing the pressurized fuel to flow from the cut-off port (2) into the pump housing. This reduces the pressure of the fuel, cutting off the injection to the nozzle.

Equalizing Stroke

As the plunger continues to rotate after the end of the delivery, figure 7-10, the equalizing slit (1) comes into line with the outlet passage. This causes the pressure in the passage to the delivery valve to be restored to the feed pressure.

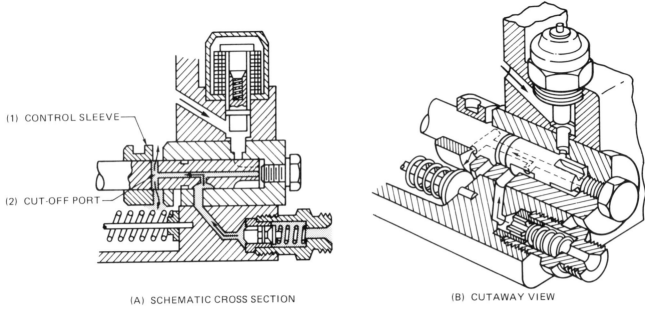

(1) CONTROL SLEEVE

(2) CUT-OFF PORT

(A) SCHEMATIC CROSS SECTION

(B) CUTAWAY VIEW

Figure 7-9 End of delivery *(Courtesy of General Motors Product Service Training)*

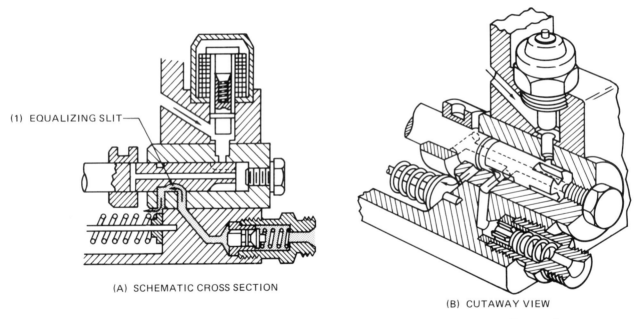

(1) EQUALIZING SLIT

(A) SCHEMATIC CROSS SECTION

(B) CUTAWAY VIEW

Figure 7-10 Equalizing stroke *(Courtesy of General Motors Product Service Training)*

The equalizing stroke maintains stable injection performance by preventing pressure interference among the fuel supplied to the various delivery valves. The equalizing slit is 180 degrees from the distributing slit, so that equalization takes place after one crankshaft revolution.

Reverse-Rotation Prevention

Assuming the plunger rotates in the normal direction, figure 7-11, during the intake stroke (left), the intake port opens causing fuel to enter the plunger. During the subsequent stroke (right), the intake port closes, and fuel is forced through the distributing slit on the plunger into one of the outlet passages.

However, if the plunger rotates in the reverse direction during the injection stroke (right), the intake port opens up, preventing the pressure from rising. As a result, injection does not take place and the engine stops.

Injection Quantity Control

Fuel quantity is controlled by sliding the control sleeve (1), figure 7-12. Shifting the control sleeve causes the *effective stroke* (plunger stroke from the beginning to the end of fuel feed) to change.

When the control sleeve (1) is moved to the left, the effective stroke decreases, reducing the injection quantity.

Conversely, when the control sleeve (1) is moved to the right, the injection quantity increases. In this way, the beginning of injection is always constant, while the end of injection is varied by moving the control sleeve to regulate the overall injection quantity.

DELIVERY VALVE

While the plunger is pressurizing the fuel, the pressure increases and overcomes the force of the delivery valve spring (1), figure 7-13a. The spring is thus compressed, causing the valve (2) to become unseated. The fuel then flows through the injection pipe to the nozzle holder, and is injected from the tip of the nozzle into the combustion chamber of the engine. Next, the plunger advances to the right while rotating, and uncovers the cut-off port,

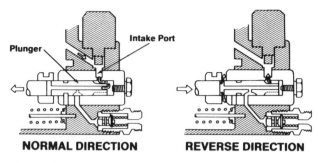

Figure 7-11 Reverse-rotation prevention (*Courtesy of General Motors Product Service Training*)

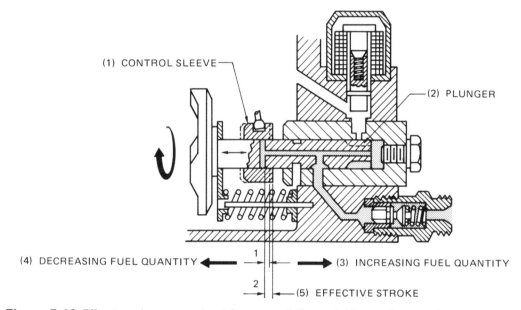

Figure 7-12 Effective plunger stroke (*Courtesy of General Motors Product Service Training*)

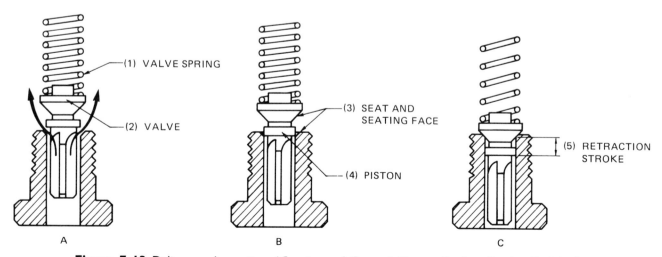

Figure 7-13 Delivery valve action (*Courtesy of General Motors Product Service Training*)

which reduces the pressure holding the delivery valve open. The pressure in the line plus the spring pressure act to close the delivery valve, figure 7-13b. The valve (2) is forced back onto the seat (3) by the spring (1), preventing reverse flow of fuel oil.

This valve incorporates a piston (4). The piston (4) land closed fuel flow at the seat bore. Further movement, as the valve is pushed to its seat, reduces the pressure in the injection pipe, cleanly cutting off injection and preventing dribble, figure 7-13c. The final step is the closing of the delivery valve, which increases the available line volume, thereby decreasing the pressure.

MIN-MAX GOVERNOR

A mechanical governor consists of a centrifugal speed regulator, employing flyweights. The flyweight holder is integral with a gear mounted on the governor shaft, and turned by the gear on the drive shaft. The flyweight holder contains four flyweights acted on by centrifugal force, causing them to open outward and push against the governor sleeve.

The governor lever assembly is supported by two pivot pins that thread from the outside of the housing (M_1), figure 7-14. The extreme outer part of this lever is called the corrector lever. The bottom part of the corrector lever is pushed by two support springs, hence the lever is pushed against the full-load adjusting screw, pivoted about point M_1. The governor lever assembly is an integrated assembly. A tension lever and starting lever are fitted to the inside of the corrector lever, and supported by the common shaft (M_2) at its bottom part.

Between the tension lever and starting lever are a starting spring (leaf spring) and start idling spring.

At the bottom of the governor lever assembly is a ball head pin that is fitted into the pinhole of the control sleeve to transmit the motion of the governor sleeve. In addition, the control sleeve is guided on the plunger to control the injection quantity.

In the control shaft assembly, figure 7-15, the governor spring and partial-load spring are fitted in the yoke under

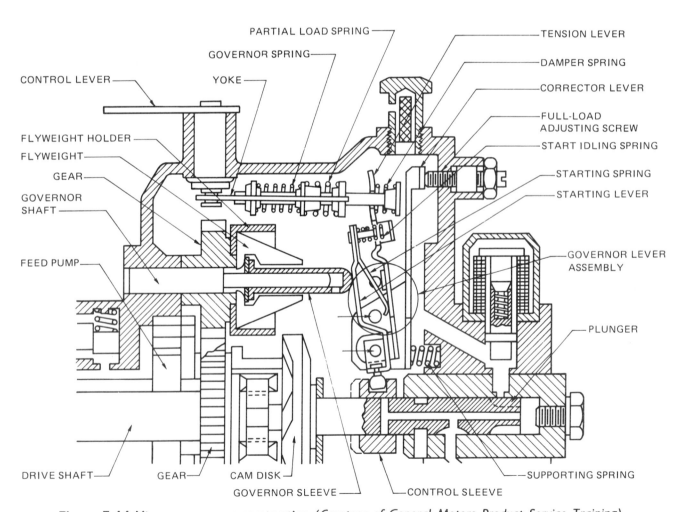

Figure 7-14 Min-max governor construction (*Courtesy of General Motors Product Service Training*)

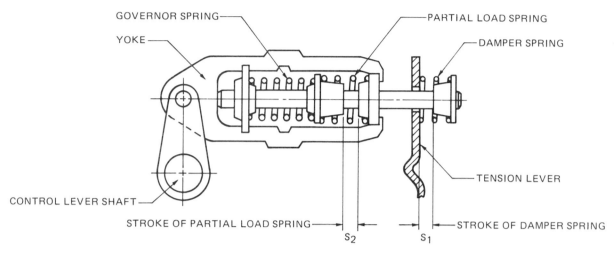

Figure 7-15 Control shaft assembly *(Courtesy of General Motors Product Service Training)*

a predetermined load. The tension of these springs determines the control speed.

Figure 7-16 shows the control characteristics of the min-max all-speed governor. A detailed explanation follows.

Starting

During starting, figure 7-17, when turning the control lever toward full direction, the starting lever is pushed left under the action of the starting spring (leaf spring). At the same time, the governor sleeve is shifted to the left. The control sleeve is moved to the right, pivoted around the common shaft (M_2) of the governor lever assembly. In this way, the control sleeve is in the maximum injection quantity setting to supply additional fuel facilitating engine starting.

When the engine starts, centrifugal force moves the flyweights out, pushing the governor sleeve to the right. This pushes the starting lever until it overcomes the force of the starting spring and moves to the right.

As a result, the control sleeve shifts to the left, shortening the effective stroke and reducing the fuel supply.

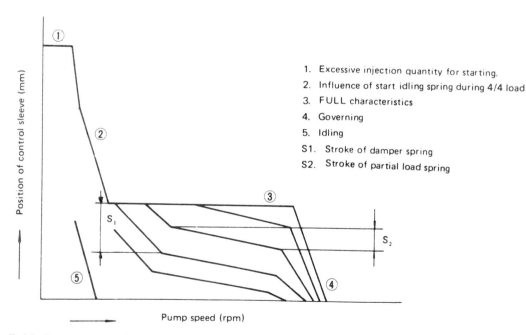

1. Excessive injection quantity for starting.
2. Influence of start idling spring during 4/4 load
3. FULL characteristics
4. Governing
5. Idling
S1. Stroke of damper spring
S2. Stroke of partial load spring

Figure 7-16 Governor speed control characteristics *(Courtesy of General Motors Product Service Training)*

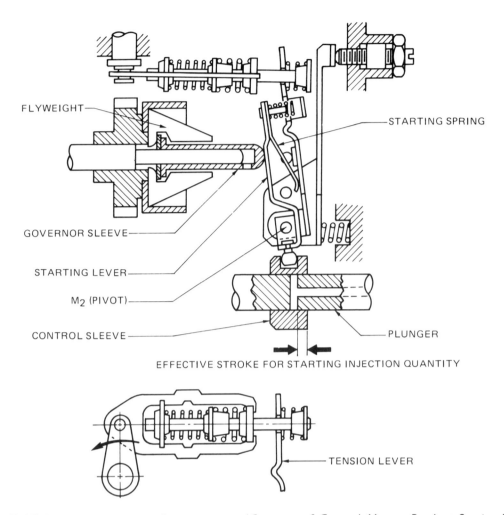

Figure 7-17 Governor operation during starting *(Courtesy of General Motors Product Service Training)*

Idling

When the engine starts up, figure 7-18, centrifugal force acts on the flyweights, causing the governor sleeve to move to the right. As a result, the starting lever and starting spring continue to be pushed until smooth idling is obtained. This happens when the combined tension force of the start idling spring and the starting spring is balanced against the centrifugal force of the flyweights.

Consequently, the control sleeve is shifted to the left, pivoted about the common shaft (M_2) of the governor lever assembly. When it reaches the idling position, the fuel supply is reduced to maintain idle speed.

Partial Loading

The control lever moves together with the accelerator pedal to which it is interlocked, figure 7-19.

If the control lever is put in any medium-speed running position, the compression force on the spring in the yoke increases, compressing the damper spring. This moves the tension lever to the left, and the control sleeve to the right [pivot at (M_2)]. As a result, the injection quantity increases, and the engine speed rises.

At the same time, the centrifugal force on the flyweights increases, causing the governor sleeve to be pushed to the right. Consequently, the starting lever is pushed to the right until it touches the raised part of the tension lever (point a in the diagram). It then moves together with the tension lever.

As a result, the centrifugal force on the flyweights causes the partial-load spring to compress until the compression force balances the centrifugal force. The tension lever shifts to the right by the amount of movement (stroke) at (S_2) in figure 7-19. At this point, the damper spring is already in a nonoperational condition.

As a result of the above, the control sleeve moves to the left, pivoted about the common shaft (M_2) of the governor lever assembly, and the fuel injection quantity is reduced.

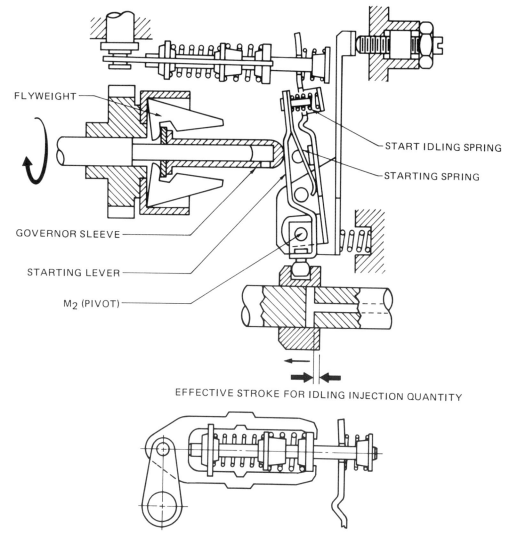

FLYWEIGHT

START IDLING SPRING

STARTING SPRING

GOVERNOR SLEEVE

STARTING LEVER

M₂ (PIVOT)

EFFECTIVE STROKE FOR IDLING INJECTION QUANTITY

Figure 7-18 Governor operation during idle *(Courtesy of General Motors Product Service Training)*

Thus, during medium- or low-speed running, the centrifugal force on the flyweights balance with the combined force of the damper spring and partial-load spring.

Full-Load Maximum Speed

The control lever is moved from a partial-load position to the full-load maximum-speed position, i.e., until it touches the external maximum-speed stop bolt, figure 7-20. The yoke is pulled completely to the left so that the compression force acting on the partial-load spring in the yoke increases. This causes the tension lever to be drawn to the left until it touches the stopper pin (M₃) and comes to rest.

The partial-load spring is in a nonoperational condition (stroke [S₂] in figure 7-19 will disappear). The governor spring is now being compressed to act against the governor.

Consequently, the centrifugal force on the flyweights balances against the governor spring at this position, enabling full load and maximum speed to be obtained.

If the full-load adjusting screw is screwed inward, the collector lever turns counterclockwise about (M₁). The control sleeve moves toward the right to increase the fuel supply.

(M₁) consists of two pivot pins that support the corrector lever on both left and right sides. They support the governor lever assembly from the outside of the housing.

No-Load Maximum Speed

When the engine speed is raised past the full-load maximum speed, the centrifugal force on the flyweights becomes maximum, figure 7-21. This overcomes the tension of the yoke spring that pulls the tension lever. The

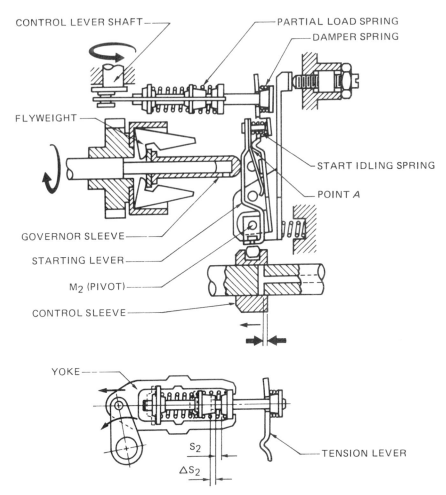

Figure 7-19 Governor operation during partial loading (*Courtesy of General Motors Product Service Training*)

governor sleeve moves to the right, pushes the tension lever, and compresses the governor spring. The centrifugal force of flyweights then balances with the governor spring tension to obtain the no-load maximum position.

The tension lever pivots at shaft (M_2) to move the control sleeve. The control sleeve moves to the left until the cut-off part of the plunger opens into the pump chamber, preventing the engine speed from rising.

INJECTION TIMING ADVANCE

Diesel injection timing is advanced by a hydraulic piston in the injection pump, figure 7-22. As engine speed increases, fuel pressure from the vane pump also increases. Vane pump pressure pushes the injection advance piston to the left against the spring, causing the roller housing to turn slightly.

Since the cam plate is turning in the opposite direction, the "ramps" on the cam plate engage the roller sooner, whenever the injection advance piston moves to the left.

This causes the distributor plunger to begin injection sooner.

The injection timing advance piston is located in the bottom of the injection pump body.

COLD START DEVICE (CSD) (MANUAL)

The only cold start and warmup device necessary for the diesel fuel system is a control that advances injection timing at idle and during low-speed running.

A lever turns a cam that pushes the hydraulic piston to the left. This advances injection timing about 5 degrees, figure 7-23.

This injection advance provides more time for the fuel to burn, which improves performance and prevents smoking during cold starts and warmup.

The cold start cam does not advance the complete range of injection timing. Above 2200 RPM the piston operates normally and does not contact the cam.

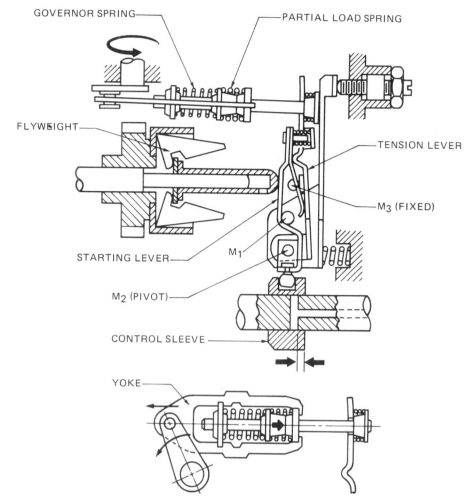

Figure 7-20 Governor operation at full-load maximum speed (*Courtesy of General Motors Product Service Training*)

COLD START DEVICE (CSD) (AUTOMATIC)

There are different types of automatic CSD available for the VE injection pump. One popular type is the vacuum-operated CSD. This type uses a vacuum pump, water thermo valve, and a dual stage diaphragm located on the injection pump, see figure 7-24. The thermo valve, figure 7-25, applies vacuum to the diaphragm(s), depending on coolant temperature. Figure 7-26 shows which diaphragms have vacuum applied at various temperatures.

ANEROID COMPENSATOR

The VE-type pump can be equipped with an aneroid compensator, figure 7-27.

The aneroid compensator is mounted on the top of the VE pump proper and is designed to control the fuel injection to match the altitude (height above sea level). This prevents deterioration of the exhaust emissions.

As the altitude increases, the atmospheric pressure decreases. The bellows (1) force eventually becomes larger than the spring (2) setting. It expands and pushes the adjusting rod (3) downward.

The pin (4) contacting the adjusting rod pushes the top end of the lever (5) to the left. This motion is transmitted via pivot (A), causing the bottom end of the lever to push the tension lever (6) to the right. This motion is transmitted about the pivot (B) of the tension lever, causing the control sleeve (7) to move in the fuel decrease direction, which is to the left.

BOOST PRESSURE ENRICHMENT DEVICE

The boost pressure enrichment device is added to the fuel injection pump on engines equipped with a turbo-

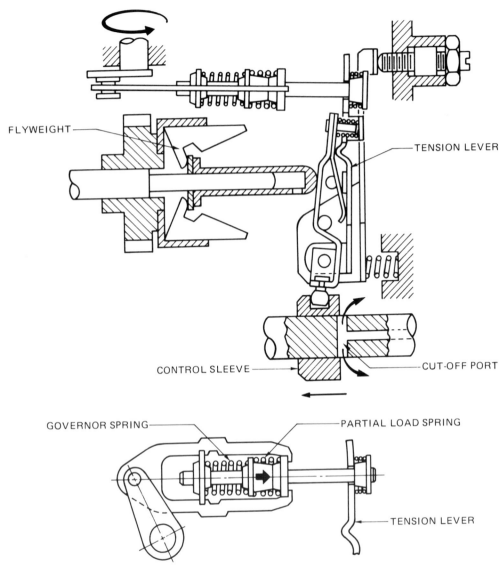

FLYWEIGHT

TENSION LEVER

CONTROL SLEEVE

CUT-OFF PORT

GOVERNOR SPRING

PARTIAL LOAD SPRING

TENSION LEVER

Figure 7-21 Governor operation at no-load maximum speed *(Courtesy of General Motors Product Service Training)*

charger. Under boost pressure from the turbocharger, more air is supplied to the cylinders. Hence, more fuel must also be supplied. In figure 7-28, note the position of the diaphragm, diaphragm pushrod, bell crank, and control ring. As boost pressure increases, the diaphragm is forced downward, changing the position of the bell crank, figure 7-29. The control sleeve (ring) moves further to the right, increasing the effective (working) stroke and the amount of fuel that is injected into the cylinder.

FUEL-CUT SOLENOID

The fuel-cut solenoid is controlled by the ignition switch, figure 7-30. It is used to open or close the fuel intake passage (2) to the intake port (3).

When the engine is running, the fuel-cut solenoid valve is open, causing the armature connected to the valve to draw in, and the passage to the plunger inlet hole on the head to open. Conversely, when the engine is off, the solenoid is de-energized. As a result, the valve is forced against the seat by the spring, cutting off the fuel supply to the plunger.

SERVICING THE VE INJECTION PUMP

All the precautions given in the previous chapter hold true for this pump. Briefly stated, they are:

● Cleanliness.
● All fasteners torqued to specifications.

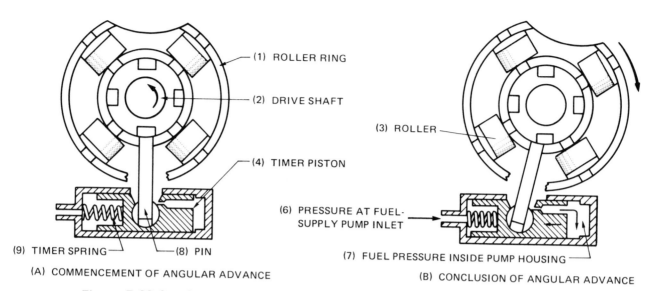

Figure 7-22 Speed timer operation *(Courtesy of General Motors Product Service Training)*

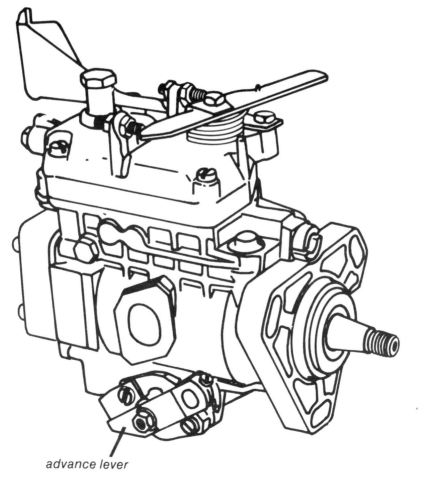

advance lever

Figure 7-23 Manual cold start advance lever *(Courtesy of Volkswagen of America Inc.)*

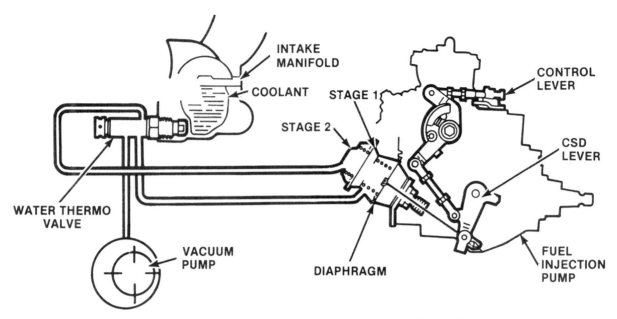

Figure 7-24 Vacuum-operated CSD *(Courtesy of Ford Motor Company)*

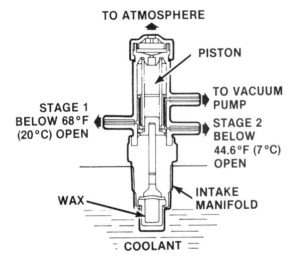

Figure 7-25 Thermo valve *(Courtesy of Ford Motor Company)*

- Use the manufacturer's special tools.
- This pump is used by several manufacturers and is manufactured under license by others. Always consult the service manual specific to the vehicle you are working on. It will state how far you can go on pump service and adjustments.
- The following procedures are meant only as a guide to explain the procedures. Use the detailed service procedures given in the service manual.

Pump Removal and Installation

This task requires the following:

- appropriate hand tools
- manufacturer's special tools
- a marking pen or scribing tool
- manufacturer's service manual

THERMO VALVE TEMPERATURE	DIAPHRAGM FUNCTION	TIMING ADVANCE
Below 45°F (7°C)	Stage 1 + Stage 2	8 degrees
45–68°F (7–20°C)	Stage 1	5 degrees
Above 68°F (20°C)	None	None (= TDC)

Figure 7-26 CSD diaphragm operation *(Courtesy of Ford Motor Company)*

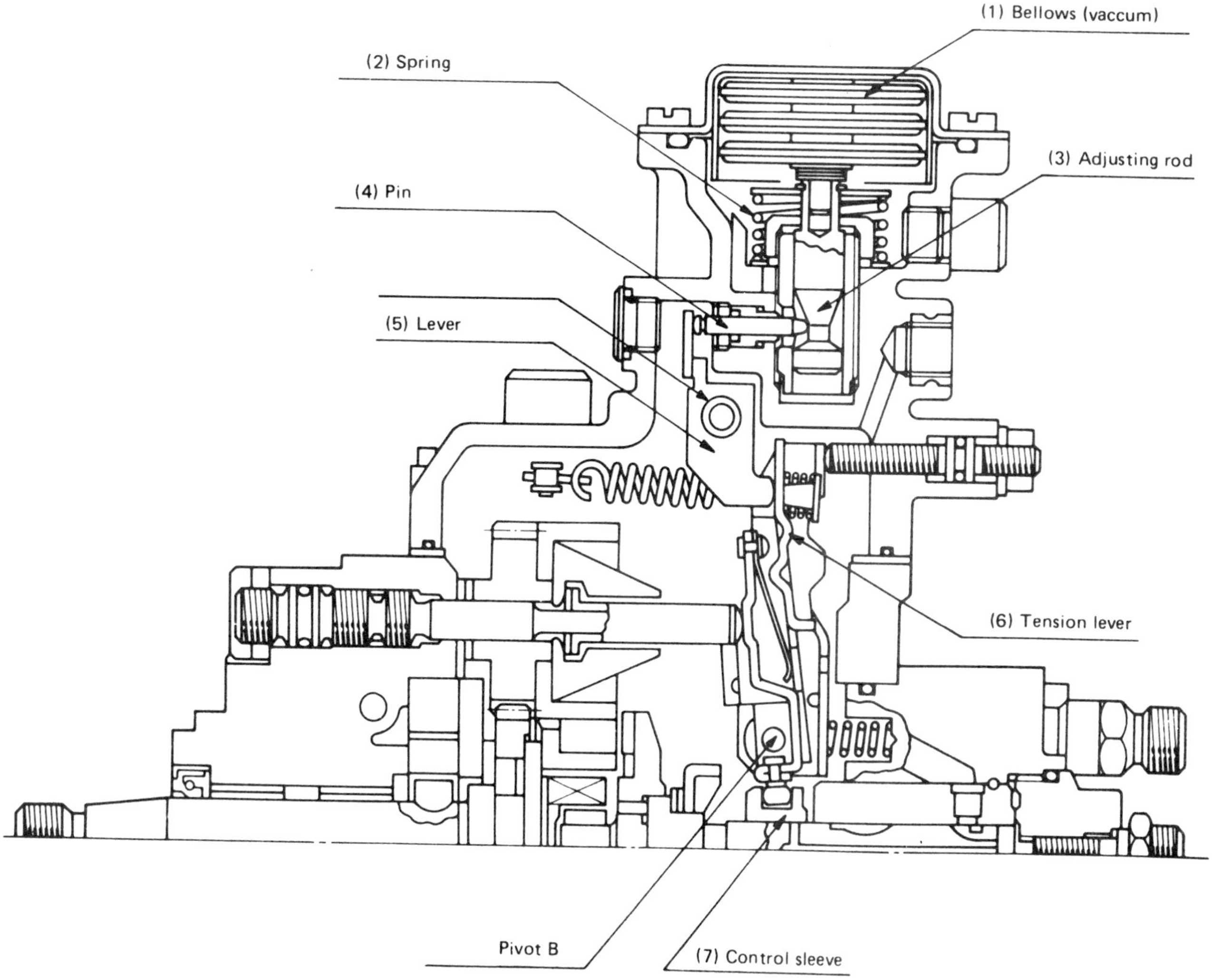

Figure 7-27 Aneroid compensator (*Courtesy of General Motors Product Service Training*)

Removal. Carefully clean and wipe the injection pump and surrounding area.

CAUTION: Mark the injection pump mounting flange to the timing case or bracket.

This will help preserve injection pump timing.

If the injection pump is belt driven, remove the belt cover.

CAUTION: Mark the belt to the injection pump gear, camshaft gear, and crankshaft gear.

This will help ensure that all components are timed properly.

Disconnect the throttle linkage and fuel cut-off solenoid wire.

Disconnect and remove the fuel injection lines. Note which cylinder each fuel line goes to. Cap all exposed openings.

Disconnect the fuel lines.

Remove the injection pump from the pump gear. Special tools may be required to perform this task. Be careful not to drop any parts inside the timing case.

Remove the injection pump's attaching nuts and bolts, and remove the pump.

Installation. Attach the pump to the engine and line up the timing marks. Tighten the attaching nuts and bolts.

Connect the injection pump to the pump gear.

Connect all fuel lines and injection lines.

Connect the throttle linkage and fuel cut-off solenoid wire. Check the throttle linkage to be sure it does not bind or stick.

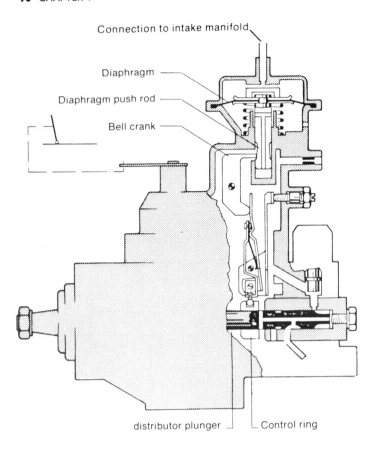

Connection to intake manifold

Diaphragm

Diaphragm push rod

Bell crank

distributor plunger   Control ring

Figure 7-28 Boost pressure enrichment device (*Courtesy of Volkswagen of America Inc.*)

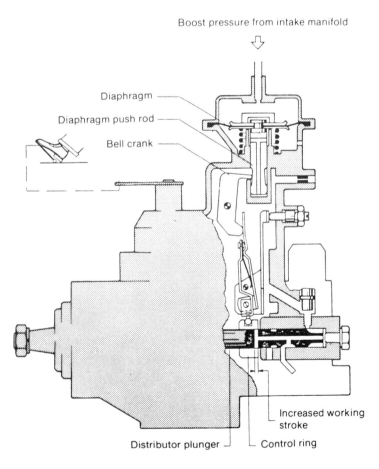

Boost pressure from intake manifold

Diaphragm

Diaphragm push rod

Bell crank

Increased working stroke

Distributor plunger   Control ring

Figure 7-29 Diaphragm pushed downward by boost pressure, increasing effective stroke (*Courtesy of Volkswagen of America Inc.*)

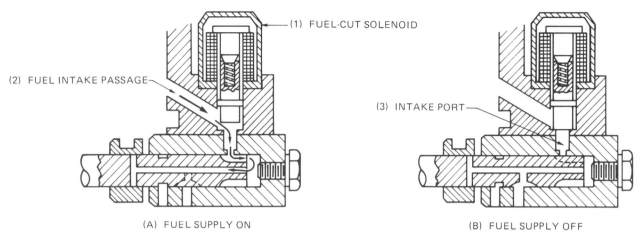

(1) FUEL-CUT SOLENOID

(2) FUEL INTAKE PASSAGE

(3) INTAKE PORT

(A) FUEL SUPPLY ON

(B) FUEL SUPPLY OFF

Figure 7-30 Fuel cut-off solenoid operation *(Courtesy of General Motors Product Service Training)*

Attach the timing belt and be sure that all timing marks are lined up.

Adjust timing belt tension. Bleed air from the injection pump according to manufacturer's instructions.

Idle Speed Adjustment

This task requires the following:

- appropriate hand tools
- special tachometer
- manufacturer's service manual

Since the VE pump is used by several engine manufacturers, there are different methods and tachometers used to set idle speed. It may be necessary to use the magnetic probe-type, photoelectric-type, or the vibration sensor-type to measure engine RPM.

The photoelectric-type shines a beam of light onto reflective tape that is mounted on the crankshaft pulley, figure 7-31. The beam of light reflects off the tape back to a sensor mounted in the tachometer. The engine RPM measurement is then read off a meter. It is important to

be sure the pulley flange is clean and that the tachometer does not pick up stray reflections.

The vibration sensor mounts on top of the valve cover, figure 7-32. RPM signals are generated by the sensor and sent to the tachometer where engine RPM measurements are read.

Prepare the vehicle according to manufacturer's instructions. This usually means with the engine fully warmed up and all accessories off.

Check idle speed. To adjust idle speed, loosen the idle speed lock nut and turn the idle speed bolt to achieve the desired reading, figure 7-33. Tighten the locking nut and recheck idle speed.

High-Speed Adjustment

Be sure the manufacturer states that this adjustment can be done; otherwise, do not do it.

This task requires the same tools used in setting idle speed.

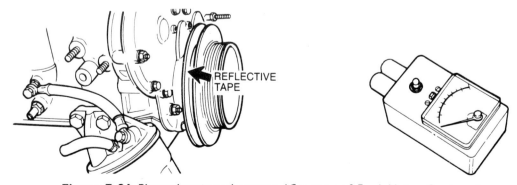

REFLECTIVE TAPE

Figure 7-31 Photoelectric tachometer *(Courtesy of Ford Motor Company)*

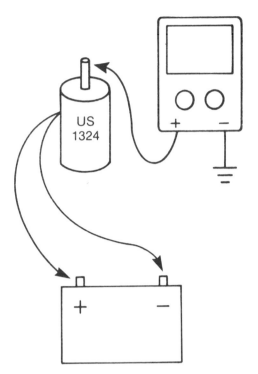

Figure 7-32 Vibration sensor tachometer *(Courtesy of Volkswagen of America Inc.)*

CAUTION: Be certain that the vehicle has the parking brake set and the wheels blocked.

With the tachometer connected and the engine at operating temperature, briefly accelerate the engine to full-load position and note the tachometer reading. If necessary, adjust the high-speed screw to manufacturer's

specifications (see figure 7-34). Do not exceed the specified speed or engine damage could result.

Injection Pump Timing

This task requires the following:

● appropriate hand tools
● manufacturer's special tools (static timing adapter, dial indicator)
● manufacturer's service manual

VE injection pump timing is performed with the engine off. Despite the fact that it is used by several engine manufacturers, the timing procedure is basically the same. What must be measured is the amount of injection pump plunger travel at a specific point (usually engine TDC), figure 7-35.

Prepare the vehicle according to the service manual. Check to be sure valve timing and timing belt tension are properly set.

Remove the screw plug in the hydraulic head. Mount the static timing adapter in the pump, figure 7-36.

Rotate the crankshaft to the specified position.

Mount the dial indicator and set it to the specified reading.

Rotate the crankshaft (usually counterclockwise) until the dial indicator stops moving. Zero gauge the pointer.

Rotate the crankshaft clockwise to the specified position and note the dial indicator reading, figure 7-37. The reading should be within specifications. If it is less than specified, timing is retarded. Greater than specified readings indicate timing is advanced.

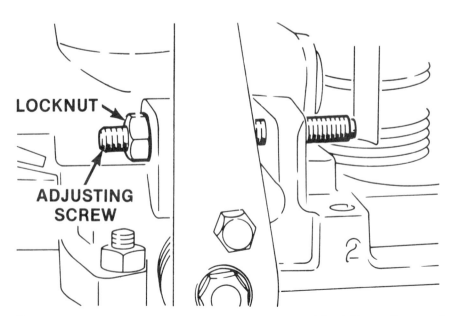

Figure 7-33 Low idle speed adjustment *(Courtesy of Ford Motor Company)*

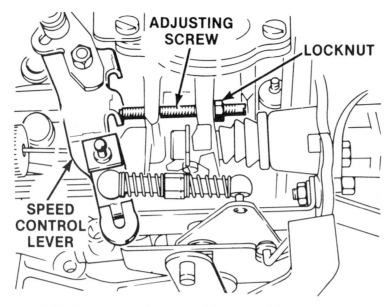

Figure 7-34 High-speed adjustment (*Courtesy of Ford Motor Company*)

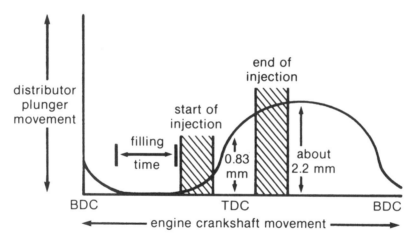

Figure 7-35 Plunger movement versus engine crankshaft movement (*Courtesy of Volkswagen of America Inc.*)

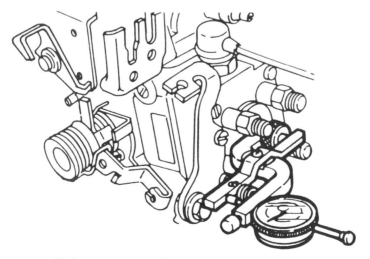

Figure 7-36 Timing gauge (dial indicator) installed on injection pump (*Courtesy of Ford Motor Company*)

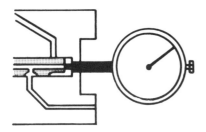

Figure 7-37 Dial indicator measuring plunger stroke (*Courtesy of Volkswagen of America Inc.*)

If it is necessary to adjust timing, loosen the pump-mounting nuts (bolts) and rotate the pump to the desired reading. Tighten the nuts (bolts) and check timing.

Remove special tools and install the screw plug. Install any other components removed. Check idle speed; adjust if necessary.

SUMMARY

The Robert Bosch VE injection pump, made by a variety of manufacturers under license, is primarily found on small, high-speed diesel engines. It can be tailored to meet specific engine needs.

This pump uses a single plunger that rotates and reciprocates at the same time. The amount of fuel injected is determined by the effective stroke of the plunger.

Care is needed when performing service on this pump. Do only the service stated by the manufacturer.

CHAPTER 7 QUESTIONS

1. Identify the main components of the Robert Bosch VE distributor pump.
2. Describe and trace the fuel flow with the aid of a diagram.
3. Explain the operation of the transfer pump and regulator assembly.
4. Explain the intake and injection strokes.
5. Describe the purpose of the equalizing stroke.
6. Explain how fuel quantity is controlled.
7. Explain the term *effective stroke.*
8. Explain the operation of the delivery valve.
9. Explain the operation of the min-max governor.
10. Explain the operation of the timer mechanism.
11. Explain the operation of the cold start device (manual and automatic).
12. Explain the purpose of the aneroid compensator.
13. Explain how the fuel cut-off solenoid works.

chapter 8
The Robert Bosch
In-line Injection Pump

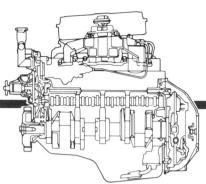

Objectives

In this chapter you will learn:

- To identify the Robert Bosch in-line injection pump
- To identify its main components
- The pattern of the fuel flow
- The components of the pumping element
- How fuel is pressurized
- How fuel is metered
- The operation of the delivery valve
- The operation of the governor
- The operation of the automatic advance mechanism
- The operation of the fuel shut-off device
- How to perform specified services

IDENTIFYING THE ROBERT BOSCH IN-LINE INJECTION PUMP

An identifying nameplate is located on the side of the injection pump. A typical code may read:

PE S 4 M 55 C 320 R S58

a. PE—in-line pump with enclosed camshaft
b. S—flange mounted
c. 4—number of high-pressure outlets
d. M—size of injection pump (note: MW is another popular size used on cars and small trucks)
e. 55—plunger diameter (5.5 mm in this case)
f. C—execution, third design change
g. R—right hand (clockwise rotation)

The remaining numbers and letters are for engineering and application information.

MAIN COMPONENTS

For the main components of the Robert Bosch in-line injection pump, refer to figures 8-1 and 8-2.

The main rotating components are the automatic advance timing unit, camshaft, and governor. The main reciprocating components are the tappets, pumping elements, and fuel feed pump.

The camshaft lobes are timed so that as the pump camshaft rotates (one-half engine speed), the tappet and plunger assembly is pushed upward. The plungers are actuated in engine-firing sequence. There is one pumping

element for each cylinder. An extra camshaft lobe is provided to actuate the fuel feed pump.

The control rack rotates the plungers, thereby controlling the amount of fuel that is injected.

Above the pumping element, inside each discharge fitting, is the delivery valve and reverse flow dampening valve. The injection lines are attached to the discharge fittings.

Attached to the back of the injection pump housing is the governor assembly. Older models used a pneumatic governor; newer models use a mechanical min-max governor.

The mechanical automatic advance is located at the front of the pump and is mounted on the camshaft, which advances or retards injection timing, according to engine speed.

The upper end of the injection pump (pumping elements, etc.) is lubricated by diesel fuel. The lower end of the injection pump (camshaft, bearings, etc.) is lubricated by engine oil.

FUEL FLOW

In figure 8-3, the fuel supply (feed) pump, located on the side of the injection pump, draws fuel from the fuel tank through the tank strainer and the in-line prefilter. The fuel supply pump then pumps the fuel under low pressure to the main filter and the injection pump. On the vehicle, the main filter is mounted higher than the

FUEL INJECTION PUMP

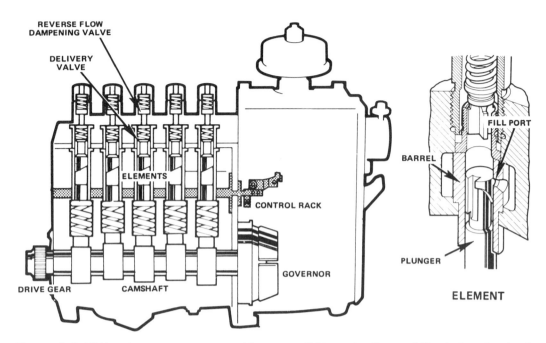

Figure 8-1 MW in-line injection pump *(Courtesy of Mercedes-Benz of North America Inc.)*

injection pump to allow air to escape through the overflow circuit. The fuel supply pump supplies more fuel than needed for injection. This helps prevent the formation of air bubbles and provides a constant delivery pressure to the injection pump. Excess fuel is returned to the fuel tank via the overflow circuit.

The pumping elements receive the low-pressure fuel and deliver a metered amount of fuel to each nozzle. Any excess fuel from the nozzle is returned to the overflow circuit.

This system is usually equipped with a hand primer pump located on the fuel feed pump. This is used to bleed air and prime the fuel system without operating the engine.

FUEL INJECTION AND CONTROL

Each pumping element consists of a barrel and plunger. The plunger has two distinct motions—(1) reciprocating and (2) rotating, see figure 8-4.

The plunger stroke is produced by the camshaft and always travels the same length; therefore the plunger stroke is constant. The stroke generates the fuel pressure necessary for injection.

At the bottom of the stroke, figure 8-5, the plunger uncovers the fill port, allowing fuel to enter under fuel supply pump pressure that fills the barrel.

Delivery to the nozzle and fuel pressurization do not begin until the fill port is completely closed, figure 8-6.

The rising plunger continues delivery to the nozzle as long as the fill port is covered. This period is called the *effective stroke*, figure 8-7.

When the rising plunger uncovers the fill port, pressure is relieved and no fuel is delivered to the nozzle. This ends the effective stroke, figure 8-8.

To meter the quantity of fuel delivered, the effective stroke must be variable. Since the lower control edge is slanted in relation to the upper control edge, rotating the plunger varies the distance over which the fill port is covered, figure 8-9. The effective stroke depends on the distance from the upper control edge to the lower control edge.

Thus, metering the quantity of fuel is made possible by the slant of the lower control edge and the turning of the plunger, which varies the effective stroke, figure 8-10.

Although the plunger stroke will always be constant, the effective stroke can be varied by turning the plunger. As the plunger is turned, the distance over the fill port becomes greater and the effective stroke becomes longer, figure 8-11. The amount of fuel injected into the cylinder increases as the effective stroke becomes longer.

At full engine load, the fill port is covered for the maximum effective stroke of the plunger, and the maximum amount of fuel is delivered to the nozzle, figure 8-12.

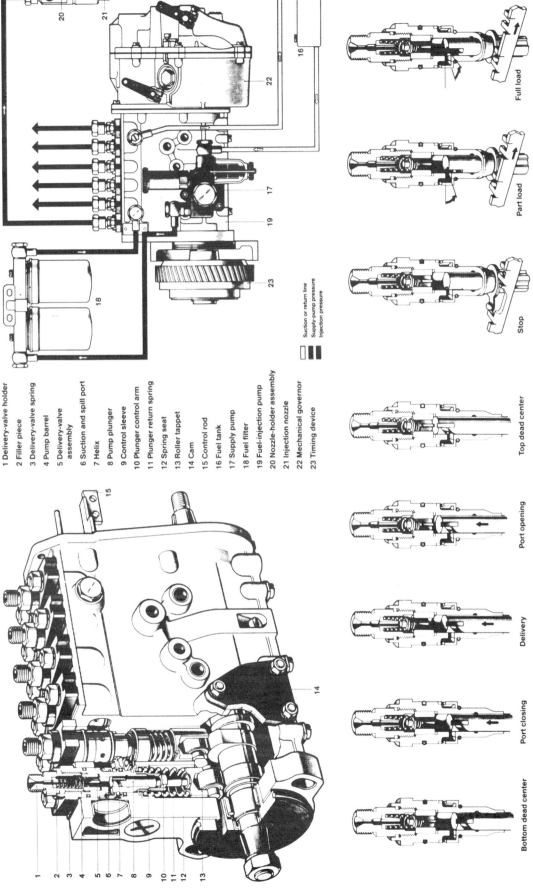

BOSCH Fuel-injection pump PES..MW
GERMANY

Robert Bosch GmbH © 1980
Training Publication VDT-U 2/104 En

1 Delivery-valve holder
2 Filler piece
3 Delivery-valve spring
4 Pump barrel
5 Delivery-valve assembly
6 Suction and spill port
7 Helix
8 Pump plunger
9 Control sleeve
10 Plunger control arm
11 Plunger return spring
12 Spring seat
13 Roller tappet
14 Cam
15 Control rod
16 Fuel tank
17 Supply pump
18 Fuel filter
19 Fuel-injection pump
20 Nozzle-holder assembly
21 Injection nozzle
22 Mechanical governor
23 Timing device

Suction or return line
Supply-pump pressure
Injection pressure

Full load

Part load

Stop

Top dead center

Port opening

Delivery

Port closing

Bottom dead center

Figure 8-2 *(Illustration only Courtesy of Robert Bosch Corporation)*

SYSTEM GENERAL LAYOUT

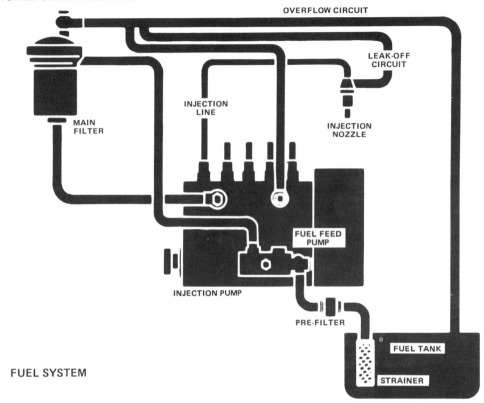

Figure 8-3 Fuel system *(Courtesy of Mercedes-Benz of North America Inc.)*

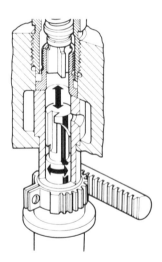
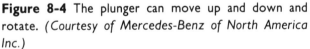

Figure 8-4 The plunger can move up and down and rotate. *(Courtesy of Mercedes-Benz of North America Inc.)*

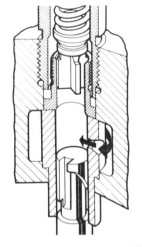

Figure 8-5 Intake stroke, plunger at BDC *(Courtesy of Mercedes-Benz of North America Inc.)*

The control rack meshes with a gear on each plunger to provide the turning motion necessary to meter the amount of fuel delivered to the nozzles. Moving the control rack forward increases the quantity of fuel delivered, while moving it backward decreases the quantity of fuel delivered to the nozzles, figure 8-13.

To shut down the engine, fuel above the plunger is not pressurized. A vertical groove on the side of the plunger is rotated, causing the groove to line up with the fill port. In this position, no fuel is pressurized; therefore no fuel is sent to the nozzles, which shuts the engine off, figure 8-14.

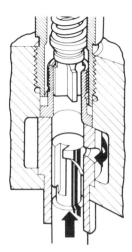

Figure 8-6 Port closing *(Courtesy of Mercedes-Benz of North America Inc.)*

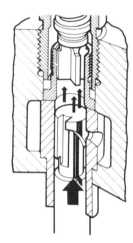

Figure 8-7 Delivery *(Courtesy of Mercedes-Benz of North America Inc.)*

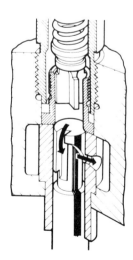

Figure 8-8 Port opening *(Courtesy of Mercedes-Benz of North America Inc.)*

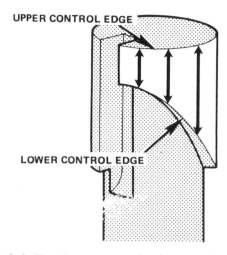

UPPER CONTROL EDGE

LOWER CONTROL EDGE

Figure 8-9 The lower control edge is cut at angle varying the distance between the upper and lower control edges. *(Courtesy of Mercedes-Benz of North America Inc.)*

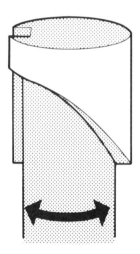

Figure 8-10 Rotating the plunger changes the relationship between the lower control edge and the fill port. *(Courtesy of Mercedes-Benz of North America Inc.)*

DELIVERY VALVE

A delivery valve is located just above each barrel and functions like the delivery valve on the Robert Bosch VE pump, figure 8-15. It allows pressure to travel to the nozzle, and maintains a residual pressure in the nozzle line.

REVERSE FLOW DAMPENING VALVE

To prevent secondary injections, a reverse flow dampening valve is incorporated above the delivery valve, figure

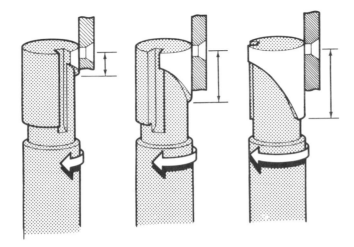

Figure 8-11 Turning the plunger varies the effective stroke. *(Courtesy of Mercedes-Benz of North America Inc.)*

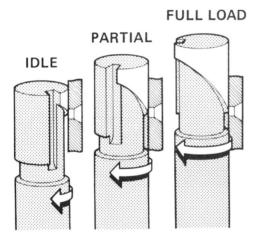

Figure 8-12 Idle—minimum effective stroke, full load— maximum effective stroke. *(Courtesy of Mercedes-Benz of North America Inc.)*

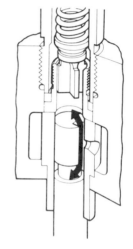

Figure 8-14 No fuel delivery. *(Courtesy of Mercedes-Benz of North America Inc.)*

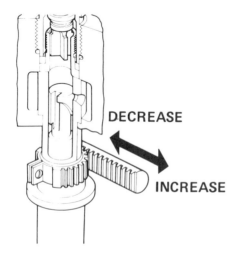

Figure 8-13 The control rack rotates the plunger. *(Courtesy of Mercedes-Benz of North America Inc.)*

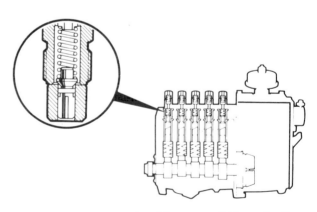

Figure 8-15 Delivery valve *(Courtesy of Mercedes-Benz of North America Inc.)*

8-16. This valve allows free flow to the nozzle and absorbs the shock wave created by the rapid closing of the nozzle.

MIN-MAX GOVERNOR

The min-max governor used on the Robert Bosch in-line pump operates in similar fashion to the min-max governors already studied. At idle and maximum engine RPM, the mechanical governor positions the control rack via internal pump linkage, figure 8-17. The accelerator pedal controls engine speed only between idle and full load as it mechanically positions the control rack. As engine speed increases, the flyweights move outward, figure 8-18. This motion is transmitted to the control rack at maximum RPM.

INJECTION TIMING

An advance unit located at the driven end of the injection pump advances timing as engine RPM increase.

As engine speed increases, flyweights move outward, advancing the position of the injection pump camshaft in relation to the drive sprocket, which is driven by the engine-timing chain, figure 8-19.

ENGINE SHUT-OFF

There are two principle methods for shutting off the fuel supply—vacuum actuated and electrically actuated.

The vacuum-actuated method uses a key switch, vacuum control unit, and a vacuum pump, figure 8-20. When the key switch is turned off, a cam in the key switch actuates a valve and vacuum is applied to the vacuum control unit. Through internal linkage, the vacuum control unit pulls the control rack to the stop position. A mechanical stop lever is provided to shut down the engine in case of malfunction.

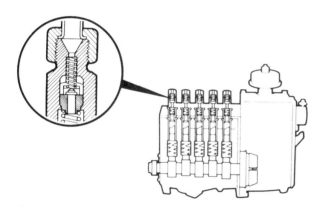

Figure 8-16 Reverse flow dampening valve (*Courtesy of Mercedes-Benz of North America Inc.*)

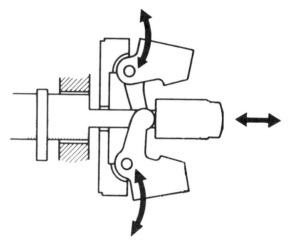

Figure 8-18 Flyweights move outward as RPMs increase. (*Courtesy of Mercedes-Benz of North America Inc.*)

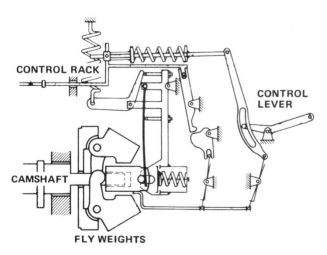

Figure 8-17 Governor (*Courtesy of Mercedes-Benz of North America Inc.*)

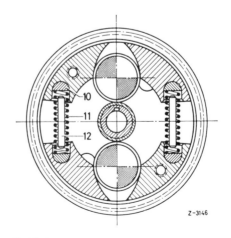

Figure 8-19 Mechanical advance mechanism. (*Courtesy of Mercedes-Benz of North America Inc.*)

SHUT-DOWN

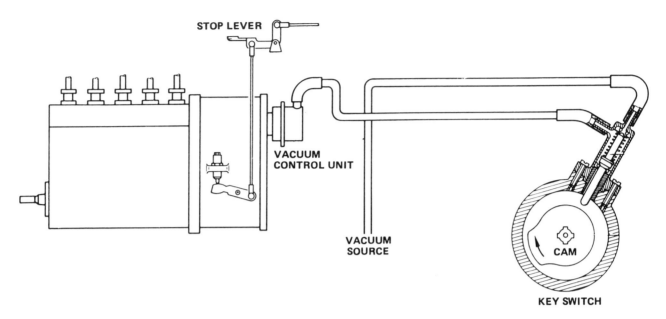

Figure 8-20 Vacuum shutdown system *(Courtesy of Mercedes-Benz of North America Inc.)*

ANEROID COMPENSATOR

The aneroid boost compensator automatically adjusts the fuel quantity injected into the cylinders, depending on the existing boost pressure or atmospheric pressure in the intake manifold. Therefore, the engine always receives the correct amount of fuel corresponding to the air charge in the cylinders, providing the most efficient operation under various conditions. The compensator is connected to the intake manifold by a pressure line.

The aneroid compensator, figure 8-21, consists of two aneroid capsules (12), a compression spring (11), a push-pull rod (10), and a gate plate (7). The push-pull rod is connected to the gate plate, and the lever (3), and the control lever (4) to the main rack (9). When boost pressure increases, the aneroid capsules (12) are compressed and, assisted by the compression spring (11), move the push-pull rod (10) in direction A. This will move the gate plate (7) within its adjusting range in direction C1, pushing the main rack (9) via the lever (3), and the control lever (4) in direction D. This increases the injected fuel quantity.

During operations at high altitudes with low engine RPM (no boost pressure), the aneroid capsules (12) expands because of reduced atmospheric pressure that forces the push-pull rod (10) against the force of the compression spring (11) in direction B. The gate plate moves in direction C2, pulling the main rack (9) by lever (3) and control

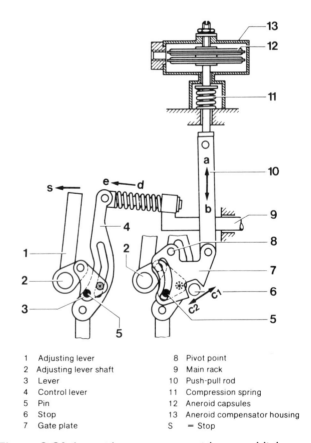

1 Adjusting lever	8 Pivot point
2 Adjusting lever shaft	9 Main rack
3 Lever	10 Push-pull rod
4 Control lever	11 Compression spring
5 Pin	12 Aneroid capsules
6 Stop	13 Aneroid compensator housing
7 Gate plate	S = Stop

Figure 8-21 Aneroid compensator with control linkage. *(Courtesy of Mercedes-Benz of North America Inc.)*

lever (4) in direction E. This decreases the injected fuel quantity.

The further the adjusting lever (1) is moved in direction S (stop), the closer the pin (5) on the lever (3) moves to the pivot point (8), gradually reducing the amount of compensatiing adjustment.

With increasing engine RPM and increased boost pressure, the aneroid compensator again increases the injected quantity of fuel according to the obtainable intake manifold pressure at the particular altitude.

SERVICING THE ROBERT BOSCH IN-LINE INJECTION PUMP

The servicing of this injection pump is performed at authorized dealers only. Typical service items done in the field are pump removal and installation, idle speed adjustment, and pump timing.

Injection pump adjustments and repair procedures vary to accommodate different engine applications. However, the following basic rules still apply:

- Clean all fittings and cap all exposed openings.
- Do only the work stated by the manufacturer to keep the pump warranty valid.
- Use the required special tools.
- Torque all fasteners to specifications.

Pump Removal and Installation

This task requires the following:

- appropriate hand tools
- manufacturer's special tools
- manufacturer's service manual

Remove the in-line injection pump by following these procedures:

1. Disconnect cables and linkages attached to the pump.
2. Disconnect fuel and lubricating lines.

CAUTION: Allow oil to drain back into the engine before disconnecting oil lines.

Cap all exposed openings.

3. With some manufacturers, the timing advance unit must be removed. Use the manufacturer's special puller to prevent damage.
4. Remove pump-mounting bolts and pull out injection pump.

Installation. Before installing the pump, it may be necessary to add oil to the injection pump. Add the amount as specified.

1. Check the control rod for proper adjustment.
2. Add required external devices from the old injection pump to the replacement.
3. Set the crankshaft in the appropriate position.
4. Install the pump so that all timing marks line up. Install the timing advance if removed.
5. Check and adjust pump timing, if necessary.
6. Tighten the mounting bolts.
7. Connect fuel lines, oil lines, linkages, and cables.
8. Bleed the fuel system.
9. Check the control linkages and adjust, if necessary.
10. Run the engine and check for leaks.
11. Check and adjust idle speed, if necessary.

Idle Speed Adjustment

This task requires the following:

- appropriate hand tools
- tachometer
- manufacturer's service manual

Adjust the idle speed by following these procedures:

1. Connect the required tachometer.
2. Check the throttle control linkage for ease of movement. Repair before going further, if necessary.
3. Run the engine to operating temperature.
4. Turn the idle speed adjuster the specified amount.
5. If equipped with a cruise control, follow the manufacturer's procedure for adjustment.
6. Check idle speed.

Injection Pump Timing

Pressurized fuel delivery on the in-line pump begins with the closing of the fill port. It also provides a point at which to set injection pump timing. By setting the engine crankshaft in a specified position, the injection pump is slightly rotated in the necessary direction until the pumping plunger closes off the fill port. This moment will be indicated by the almost complete cut-off of fuel flowing out of the no. 1 pumping element. This is the point when injection begins and is what must be adjusted.

This task requires the following:

- apppripte hand tools
- auxiliary fuel container
- overflow pipe
- manufacturer's service manual

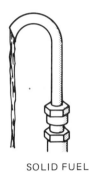 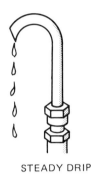 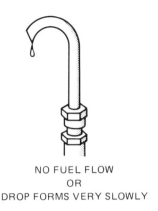

SOLID FUEL   STEADY DRIP   NO FUEL FLOW
OR
DROP FORMS VERY SLOWLY

Figure 8-22

Adjust injection pump timing by following these procedures:
1. Rotate crankshaft to specified position.
2. Remove the fuel connection fitting from the no. 1 pumping element and install overflow pipe.
3. Set control lever to maximum fuel position.
4. Attach auxiliary fuel container to the point specified on the injection pump. Note: Some manufacturers do not require the use of the auxiliary container but use the priming pump to supply fuel instead.
5. Turn crankshaft slowly in the direction of engine rotation until fuel stops flowing from the overflow pipe, figure 8-22. Note: It is normal for a drop of fuel to form 10 to 15 seconds later.
6. The crankshaft should be in the specified position. Rotate crankshaft two more turns. Fuel should stop flowing at the end of the second turn. If so, pump timing is set correctly.
7. To adjust pump timing, loosen the mounting nuts and carefully rotate the pump in the required direction. Fuel should not be flowing and the crankshaft should be in the proper position. Tighten the injection pump.
8. Remove the overflow pipe and install the fuel connection fitting with a new gasket.
9. Remove the auxiliary container.
10. Check the control lever for freedom of movement. Bleed the fuel system.
11. Run the engine and check for signs of leakage. Check idle speed.

SUMMARY

The Robert Bosch in-line injection pump uses a pumping element for each cylinder. Each element pressurizes the fuel according to engine-firing order. The amount of fuel injected is determined by the effective stoke of the plunger.

CHAPTER 8 QUESTIONS

1. Identify the main components of the Robert Bosch in-line injection pump.
2. Describe and trace the fuel flow.
3. Name the components of the pumping element.
4. Explain how fuel is pressurized.
5. Explain how fuel is metered.
6. Explain how the effective stroke is changed on the in-line pump.
7. Explain the operation of the delivery valve.
8. Explain the operation of the governor.
9. Explain the operation of the automatic advance mechanism.
10. Explain the operation of the vacuum shut-off device.

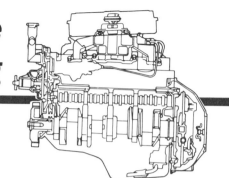

chapter 9
Injection Nozzle Operation and Testing

Objectives

In this chapter you will learn:
- How nozzles operate
- How to remove and install nozzles
- How to test nozzles

PURPOSE OF THE NOZZLE

The injection nozzle is a component of the injection system that directs a metered quantity of fuel from the injection pump into the combustion chamber. The purpose of the injection nozzle is to direct the fuel into the combustion chamber in a manner that will provide optimum engine performance with a minimum of emissions. The injection nozzle accomplishes this purpose in two ways: (1) atomizing the fuel, and (2) spreading the fuel spray in a particular pattern.

Atomization is the process of breaking down the fuel into very fine droplets. This process is necessary to mix the fuel with the compressed air, forming a vapor. For the air to mix readily with the atomized fuel, the fuel is sprayed into the combustion chamber in a particular pattern. This is called the *spray pattern*. It varies, depending on the shape of the combustion chamber and type of nozzle.

BASIC PARTS OF THE INJECTION NOZZLE

The injection nozzle is a component that can be divided into two parts: a nozzle holder and a nozzle, figure 9-1. The nozzle holder supports the nozzle in the cylinder head. The nozzle is in the lower half of the injection nozzle, which contains the parts necessary to allow and prevent fuel flow. The nozzle is the part that directs the fuel into the combustion chamber. It also contains a valve and seat that prevents fuel from flowing when the valve is held against the seat. When the valve is pushed off its seat, the fuel can exit the nozzle.

Note that some manufacturers call the nozzle and holder a fuel injector. Specifically, a fuel injector receives fuel at relatively low pressure and boosts it to a high pressure within the injector, injecting fuel into the cylinder. Examples of this would be the injectors found on Cummins

or Detroit Diesel engines. An injection nozzle receives fuel already under high (injection) pressure and does not boost pressure on its own. Because of this difference and the fuel systems used on cars and small trucks, the term "nozzle" or "injection nozzle" is used in this text.

TYPES OF NOZZLES

All nozzles in use for automotive diesel engines are differential pressure, hydraulically operated nozzles. This hydraulic action occurs when diesel fuel sent under high pressure by the injection pump overcomes spring pressure. When this happens, the nozzle valve opens, allowing the fuel to exit the nozzle. When the fuel pressure drops, the spring closes the nozzle valve, cutting off fuel flow.

Nozzles can be classified into two basic groups—inward opening and outward opening. In the inward-opening nozzle, the nozzle valve moves up into the nozzle body, figure 9-

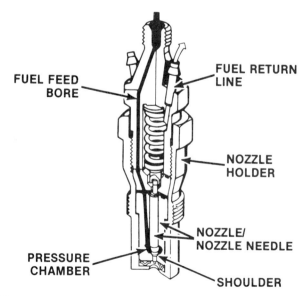

Figure 9-1 Nozzle holder and nozzle (*Courtesy of Ford Motor Company*)

2. In the outward-opening nozzle, the nozzle valve moves out, away from the nozzle body, figure 9-3. There are several variations of these two basic types of injection nozzles.

Inward-Opening Nozzles

Hole-type nozzles contain holes in the tip of the nozzle. The number and size of these holes dictate what shape the spray pattern will be for proper combustion, figure 9-4. Pintle-type nozzles use a tapered valve that seats in a single orifice in the valve body, figure 9-5. This produces a cone-shaped spray pattern.

Outward-Opening Nozzles

Poppet-type nozzles use a tapered valve that seats in a single orifice in the valve body. The poppet nozzle valve moves outward, producing a fine cone-shaped mist, figure 9-6.

OPERATION OF THE ROOSA-MASTER PENCIL-TYPE NOZZLE

The Pencil nozzle is a closed-end (nozzle valve does not project through an opening in the nozzle tip), differential pressure, hydraulically operated, hole-type nozzle, figure 9-7. The nozzle body incorporates the inlet fitting, tip, and valve guide. An edge-type filter is used to break up water droplets. The inward-opening valve is spring loaded and controlled by the pressure-adjusting screw and lift-adjusting screw, which are secured by locknuts. A nylon seal beneath the inlet fitting "banjo" prevents leakage of engine compression while a Teflon carbon dam prevents carbon accumulation in the cylinder head bore.

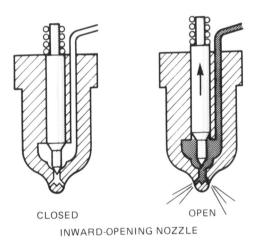

CLOSED   OPEN

INWARD-OPENING NOZZLE

Figure 9-2 Inward-opening nozzle *(Courtesy of John Deere & Company)*

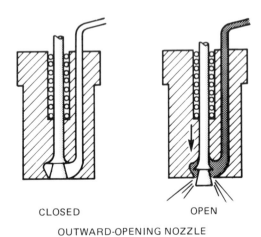

CLOSED   OPEN

OUTWARD-OPENING NOZZLE

Figure 9-3 Outward-opening nozzle *(Courtesy of John Deere & Company)*

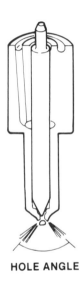

HOLE ANGLE

Figure 9-4 Number and size of holes determine the spray pattern. *(Courtesy of Ford Motor Company)*

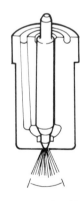

ANGLE OF SPRAY

Figure 9-5 Inward-opening pintle nozzle delivers a cone-shaped pattern. *(Courtesy of Ford Motor Company)*

Fuel Flow

Metered fuel, under pressure from the injection pump, flows through the inlet, the edge filter, and around the valve, filling the nozzle body. When the pressure acting on the differential area at the tip overcomes the spring force of the pressure-adjusting spring, the nozzle valve lifts off its seat. As the valve rises to its predetermined lift height, high-pressure fuel is allowed to flow through the spray orifices in the tip. When delivery to the nozzle ends and line pressure drops below the present nozzle-opening pressure, the spring returns the valve to its seat.

Between injections, positive sealing is maintained by the interference angle, which results in line contact between the valve and its seat.

During injection, a small amount of fuel leaks through the clearance between the nozzle valve and its guide, lubricating and cooling all moving parts. The fuel flows through a leak-off boot at the top of the nozzle body and returns to the fuel tank.

OPERATION OF THE PINTLE-TYPE NOZZLE (ROBERT BOSCH, KIKI, NIPPONDENSO)

The pintle-type nozzle is an open-end, differential pressure, hydraulically operated nozzle. It is composed of an upper half and a lower half that is threaded together, figure 9-8. The upper half houses a spring, an adjustment shim, and provides fittings for the injection pump line and return fuel line. The lower half houses the nozzle and nozzle valve (needle). The nozzle is threaded into the cylinder head and seats against a heat shield. The heat shield acts as an insulator and prevents compression loss. A variation on this design is called the center hole in

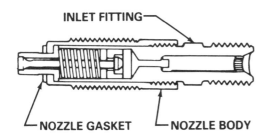

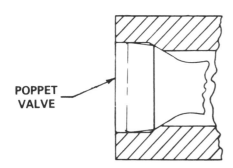

Figure 9-6 Poppet valve injection nozzle (*Courtesy of General Motors Product Service Training*)

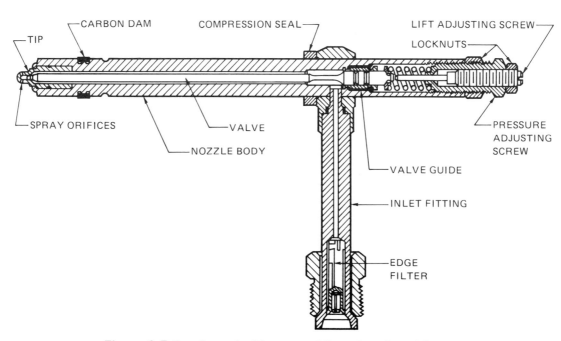

Figure 9-7 Pencil nozzle (*Courtesy of Stanadyne Diesel System*)

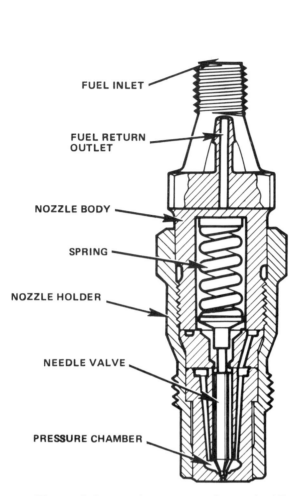

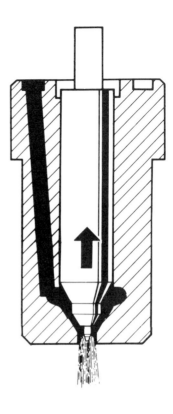

FUEL INLET

FUEL RETURN OUTLET

NOZZLE BODY

SPRING

NOZZLE HOLDER

NEEDLE VALVE

PRESSURE CHAMBER

Figure 9-8 Inward-opening pintle nozzle *(Courtesy of Mercedes-Benz of North America Inc.)*

pintle (CHIP) nozzle. The CHIP nozzle improves spray pattern for better low-speed running, figure 9-9.

Another variation on this design is called the instrumented nozzle, figure 9-10. A small coil built near the top of the nozzle body senses the nozzle needle as it lifts off its seat and the duration of the lift. An electrical signal is sent to the computer telling the computer when injection began and its duration. This information is used in two ways. First, it can be used by a computer to calculate the vehicle's fuel economy. Another computer uses the information to vary injection timing, lowering emissions, figure 9-11.

Fuel flow is nearly identical to that of the Roosa-Master pencil-type nozzle. Fuel pressure from the injection pump forces the needle up against spring pressure so that the nozzle sprays a cone-shaped mist of diesel fuel at the proper time, figure 9-12. When fuel pressure drops, the needle is pushed back into the seat. Spring pressure on most is adjusted to different shim thicknesses. The thicker the shim, the greater the spring pressure. A small amount

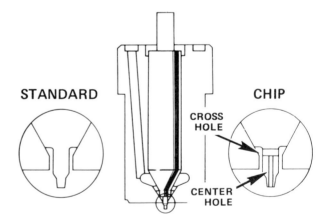

STANDARD

CHIP

CROSS HOLE

CENTER HOLE

Figure 9-9 CHIP nozzle. *(Courtesy of Mercedes-Benz of North America Inc.)*

of fuel leaks around the nozzle needle to provide cooling and lubrication. This fuel returns to the tank via the return line.

OPERATION OF THE POPPET NOZZLE (GM)

The poppet nozzle is an open-end, outward-opening nozzle, figure 9-13. It is divided into two halves—the nozzle body and the inlet fitting. The inlet fitting is threaded into the nozzle body. It houses the spring, nozzle valve, spring seat, and edge filter. The body is threaded into

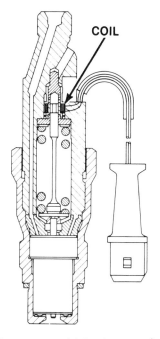

Figure 9-10 Instrumented injection nozzle (*Courtesy of Ford Motor Company*)

the cylinder head and contacts a sealing washer. The sealing washer provides a good seal to prevent compression loss.

Fuel Flow

Fuel sent by the injection pump under high pressure passes through the edge filter, then pushes against the nozzle valve, overcoming spring pressure. The nozzle valve moves outward, allowing the fuel to exit. At the lower end of the nozzle valve are a series of small, angled vanes. These aid in creating a narrow cone-shaped spray. There is no return line with this type of nozzle.

NOZZLE SERVICE AND TESTING

Nozzles are subjected to the intense heat and high pressure of the combustion chamber. Furthermore, they must provide positive seal against any fuel leakage past the nozzle tip. Nozzles can operate for long periods of time without maintenance. However, contaminated fuel, misuse, and mechanical failure can cut the nozzle's life expectancy short. For this reason nozzle servicing and testing is important.

Nozzle Removal

Dirt is the diesel engine's main enemy. Before removing any nozzles, be sure to clean the fuel connections and the area around the nozzles.

VP 20-PUMP

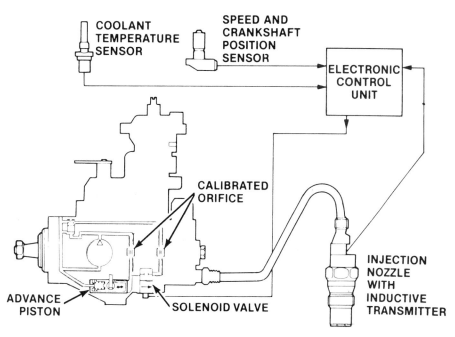

Figure 9-11 Information from the injection nozzle is used by the ECU computer to control injection timing. (*Courtesy of Ford Motor Company*)

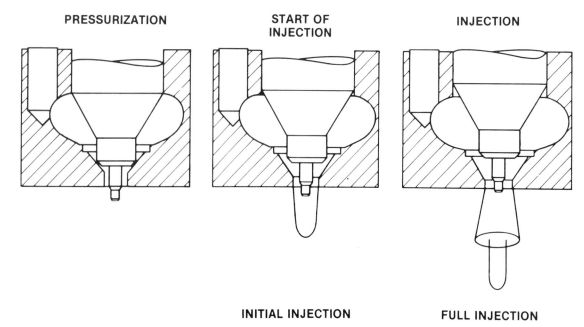

PRESSURIZATION **START OF INJECTION** **INJECTION**

INITIAL INJECTION **FULL INJECTION**

Figure 9-12 The position of the pintle (nozzle needle) greatly affects how the fuel is injected into the cylinder. In this particular figure, the pintle allows a small amount of fuel to enter the chamber. This fuel quickly ignites, while the main body of fuel follows immediately behind to continue combustion. *(Courtesy of Ford Motor Company)*

1. Carefully remove the injection lines. Use a line wrench to remove the injection line fittings and a back-up wrench on the upper hex on the nozzle body.

CAUTION: Do not bend or kink the injection lines.

2. Remove the nozzle. Use a clean deep-well socket that fits over the largest hex. The pencil-type nozzle requires a special puller. Failure to use the puller could result in distortion of the nozzle body.

3. Always cap the nozzle and lines to prevent contamination.

4. At this time, remove the sealing washer or heat shield.

Nozzle Testing

Proper checking of a nozzle requires a nozzle tester, figure 9-14. This tester can perform a series of tests on the nozzle. Three tests that are common to all nozzles are the opening pressure, spray pattern, and seat tightness tests.

Opening pressure is the point where the nozzle begins to spray the fuel. Spring pressure has a direct effect on opening pressure. The greater the spring pressure, the higher the opening pressure. Distorted or binding nozzle valves adversely affect opening pressure.

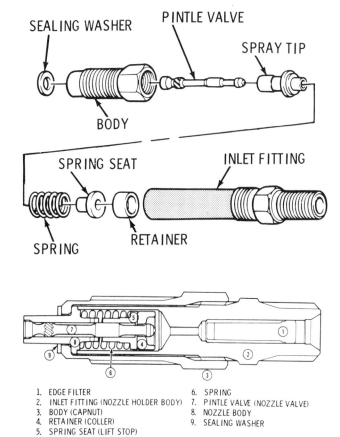

1. EDGE FILTER
2. INLET FITTING (NOZZLE HOLDER BODY)
3. BODY (CAPNUT)
4. RETAINER (COLLER)
5. SPRING SEAT (LIFT STOP)
6. SPRING
7. PINTLE VALVE (NOZZLE VALVE)
8. NOZZLE BODY
9. SEALING WASHER

Figure 9-13 Poppet injection nozzle *(Courtesy of Oldsmobile Division)*

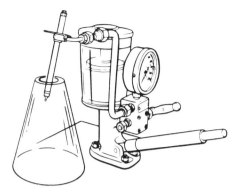

Figure 9-14 Injection nozzle tester with transparent cover over nozzle tip (*Courtesy of Stanadyne Diesel System*)

Spray pattern is the shape of the fuel as it exits the nozzle, figure 9-15. Carbon and damaged nozzle tips are two common causes of a poor spray pattern.

Seat tightness tests for fuel leakage at the nozzle tip, figure 9-16. A fuel droplet at the nozzle tip indicates a worn nozzle valve and seat.

Two more tests performed on some nozzles are the chatter and return fuel tests.

Chatter is the noise produced when the nozzle valve opens and closes rapidly. This noise indicates that the nozzle valve is moving freely in its bore.

The return fuel test determines the amount of fuel that leaks past the nozzle valve and returns to the fuel tank. This test ensures that the nozzle is being adequately cooled and lubricated.

Using the Nozzle Tester

There are two precautions that must be strictly adhered to.

1. CAUTION: Test Fuel Spray is flammable. Keep vapor away from open flames and sparks.

2. CAUTION: When testing nozzles, do not place your hands or arms near the tip of the nozzle. The high-pressure atomized fuel spray from the nozzle has sufficient penetrating power to puncture flesh and destroy tissue, which may result in blood poisoning. The nozzle tip should be enclosed in a receptacle, preferably transparent, to contain the spray.

Nozzle Test Procedure

This task requires the following:

- appropriate hand tools
- nozzle tester
- manufacturer's service manual

Perform the nozzle test procedure according to the following steps:

1. Install the nozzle on the tester,, figure 9-17.

2. Close the gauge valve. Operate the lever with rapid strokes to prime the tester and nozzle.

Open pressure test. Open the gauge valve and raise pressure slowly until the nozzle opens. Compare this reading to specifications.

Spray pattern test. 1. Close the pressure gauge valve. Operate the tester at approximately 30 strokes per minute and observe the spray pattern.

2. Compare the spray pattern to the manufacturer's description of what the spray pattern should look like, figure 9-18.

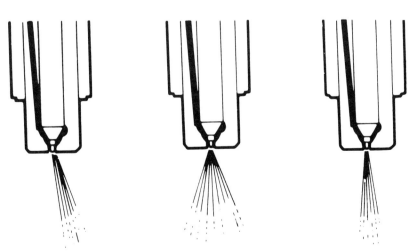

Figure 9-15 Abnormal and normal spray patterns (*Courtesy of Ford Motor Company*)

Seat tightness test. Dry the nozzle tip thoroughly. Raise the pressure and hold according to manufacturer's specifications. If any droplets have formed or are about to form, it indicates that the nozzle is faulty, figure 9-19.

Chatter test. Close the gauge valve and pump the tester lever rapidly. The nozzle should chatter.

Return fuel test. Loosen the connector nuts and reposition the nozzle tip slightly above the horizontal plane, figure 9-20. Retighten the connector nuts and raise pressure to manufacturer's specification. The nozzle should not open. Observe fluid from the nozzle return. After the first drop forms on the return fuel end of the nozzle, count the number of drops in the time period given by the manufacturer. The number of drops should be within specifications.

In case of failure. If a nozzle fails any of these tests, it must be repaired or replaced. Manufacturers may

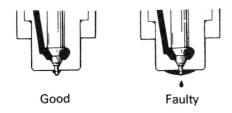

Good   Faulty

Figure 9-16 Seat tightness check

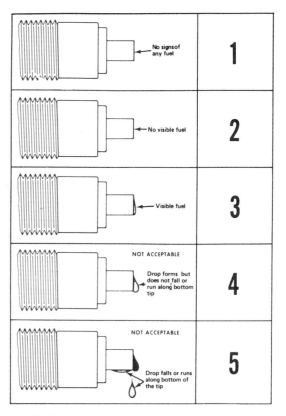

Figure 9-19 Nozzle seat tightness check (*Courtesy of Oldsmobile Division*)

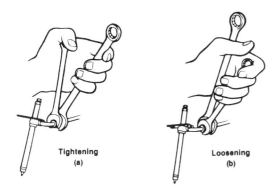

Tightening   Loosening
(a)   (b)

Figure 9-17 Installation of injection nozzle (*Courtesy of Stanadyne Diesel System*)

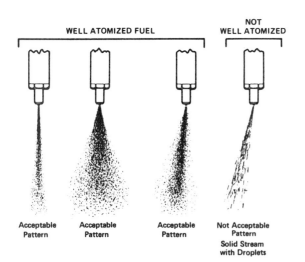

Figure 9-18 Spray patterns (*Courtesy of Ford Motor Company*)

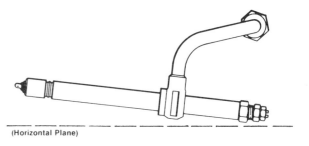

(Horizontal Plane)

Figure 9-20 Injection nozzle mounted horizontally (*Courtesy of Stanadyne Diesel System*)

recommend replacement of the nozzle. However, the nozzle may be cleaned so that it gives satisfactory performance.

Cleaning Nozzles

There are some precautions that apply to all nozzles. First, never interchange parts between nozzles. The parts are mated to each other in such close tolerances that interchanging will cause them to be too tight or too loose.

Second, never use any steel brushes or scrapers on the nozzle tip or any internal parts.

Third, a clean work area is absolutely essential.

Be sure to follow the manufacturer's procedures for disassembly of the nozzle. Note: Do not touch the nozzle mating surfaces with your fingers. The acids and dirt on your fingers can cause corrosion on the nozzle assembly.

The method for cleaning nozzles listed below requires the following:

- a wooden stick
- a brass bristle brush
- clean diesel fuel

Clean the nozzle by following these procedures:

1. Remove the carbon adhering to the nozzle needle tip with a wooden stick, figure 9-21.
2. Remove the carbon from the exterior of the nozzle body with a brass brush, figure 9-22.
3. Inspect the seat of the nozzle body and needle tip for burns or corrosion. If any of these conditions are present, replace the nozzle assembly.

4. Perform the *sinking test* using the following method:
   a. Wash the nozzle in clean diesel fuel. Note: Do not touch the nozzle mating surfaces with your fingers.
   b. Tilt the nozzle body about 60 degrees and pull the needle out one-third of its length, figure 9-23.
   c. When released, the needle should sink down into the nozzle smoothly by its own weight, figure 9-24.
   d. Repeat this test, rotating the needle slightly each time.
   e. If the needle does not sink freely, replace the nozzle assembly.

Another cleaning method uses a sonic bath cleaner. Follow the manufacturer's procedure on use of the cleaner.

Always check the nozzle on a nozzle tester after cleaning.

Nozzle Installation

1. Be sure the nozzle and nozzle area are clean. Carbon or dirt can cause improper seating that may lead to serious damage.
2. When installing the nozzle, always install a NEW sealing gasket or heat shield. This will prevent premature failure of the nozzle or compression loss.

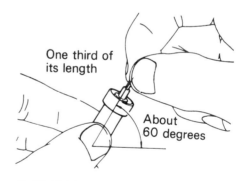

Figure 9-23 Hold nozzle body and needle as shown.

One third of its length

About 60 degrees

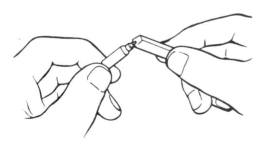

Figure 9-21 Cleaning nozzle needle

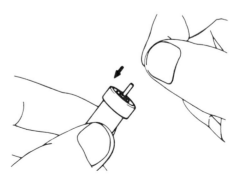

Figure 9-24 Nozzle needle should sink freely into its bore.

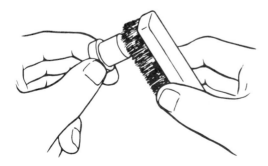

Figure 9-22 Cleaning nozzle body

3. Torque the nozzle according to manufacturer's specifications.
4. Install injection lines and secure them.
5. Run the engine to check for any fuel leakage.

SUMMARY

Nozzles are finely machined parts that are expected to atomize and spread the fuel evenly in the combustion chamber for optimum engine performance.

Nozzles are classified mainly by the direction the nozzle valve moves in relation to the nozzle body. They are further classified by the type of tip the nozzle uses.

Nozzles must be checked, cleaned, and inspected according to the manufacturer's procedures. Be sure you observe all safety precautions.

CHAPTER 9 QUESTIONS

1. What must the nozzle do to provide optimum engine performance?
2. Name two types of nozzles.
3. Explain how an inward-opening nozzle works.
4. Explain how an outward-opening nozzle works.
5. What is the main difference between a poppet nozzle and a pintle nozzle?
6. What happens to opening pressure when spring pressure is increased?
7. Name three tests common on all nozzles.
8. Why must hands or arms be kept away from the tip of the nozzle?
9. What components are faulty if the nozzle tip drips fuel?
10. Name two items that can cause poor spray pattern.

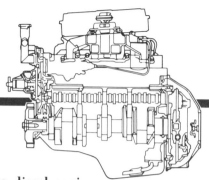

chapter 10
Diesel Engine Starting System

Objectives

In this chapter you will learn:

- Three differences between a gasoline starting system and a diesel engine starting system
- The major components of the glow plug system
- The operation of the glow plug system
- Why starting aids such as ether cannot be used
- How to test a glow plug system for proper operation

DIESEL ENGINE STARTING FACTORS

As you have learned, heat from compressed air is used for auto-ignition of the injected fuel. Anything that affects the temperature of the air within the cylinder will affect how well the engine starts. The starting characteristics of the diesel engine depend on:

- ambient temperature
- cylinder temperature
- speed of the piston
- type of combustion chamber

Cold ambient air temperatures reduce the starting efficiency of the engine. First, the colder the air, the lower its final temperature is when compressed. Second, the cold cylinder and combustion chamber walls take heat away from the air.

The speed of the piston as it travels upward on the compression stroke must be high enough to compress the air quickly and keep the compressed air's heat within the air. If the piston travels upward too slowly, the compressed air's heat is lost to the cylinder and combustion chamber's walls. The minimum cranking RPM for the diesel to start is approximately 100 RPM when cold (for gasoline engines it is 60 RPM). To crank the engine fast enough, the starter motor and battery(ies) must have sufficient power to overcome the load caused by oil viscosity when cold.

The combustion chamber type plays an important part in heat retention. Indirect injection, (IDI) engines used in automobiles and small trucks have a high ratio of combustion chamber surface to volume. Therefore, it is harder for the air to retain its heat simply because heat is conducted away through the combustion chamber walls. Also, IDI injected fuel contacts and cools the chamber walls, taking away more heat.

The goal is to build up enough heat within the compressed air to ignite the fuel. To accomplish this goal requires (a) high compression ratios, (b) a heavy-duty starting system, (c) a glow plug system, and (d) for cold climates, an engine block heater.

In chapters 2 and 3 it was discussed how high compression ratios were used in IDI engines to make up for the heat lost to the combustion chamber walls on the compression stroke. The high CR raised the cylinder temperature and reduced the amount of surface area so heat could be lost. A high-CR engine also needs more power to crank it. The CR is so high that ether starting aids should never be used! Serious engine and starter damage results because the ether mixture ignites too soon.

ENGINE-CRANKING SYSTEM

A heavy-duty starting system is needed to overcome the high CR and high oil viscosity when cold, to ensure that piston speed is high enough for the compressed air to retain its heat. The starting system used on diesel-powered cars and trucks is very similar to the systems used on gasoline-powered vehicles. However, since the engine load is much greater, greater battery capacity, larger-diameter starter motor cables, and larger starter motors are needed, figure 10-1.

Two batteries are often used to supply the needed power. The batteries are connected in parallel to maintain 12 volts and increase total amperage output, figure 10-2.

Larger-diameter starter cables are used to carry the higher-amperage requirements while keeping resistance low.

Starter motors used on diesel engines are very similar to their gasoline engine counterparts. Larger field and

	2.0L DIESEL		VS	2.0L GASOLINE
BATTERY GAPACITY (COLD CRANKING AMPS)		1000		400
CURRENT DRAW UNDER NORMAL LOAD (AMPS)		350		175
NORMAL ENGINE CRANKING (RPM)		250–350		190–260
CURRENT DRAW, NO LOAD (AMPS)		190		80
STARTER FREE-SPIN SPEED (RPM)		20,000		10,000

*ALL VALUES ARE APPROXIMATE AND WILL VARY ACCORDING TO ENVIRONMENT AND ENGINE CONDITION.

Figure 10-1 Note the differences between the same displacement diesel and gasoline engines.

armature windings are used to increase starter motor output. The gear ratio between the starter pinion gear and ring gear is reduced to decrease the starter load. On some models, manufacturers use gear reduction starters to reduce power and load requirements, figure 10-3.

Again, the starter circuit is virtually identical to the comparable gasoline-powered model, but with higher current requirements.

GLOW PLUG SYSTEMS

Glow plugs are used to preheat the combustion chamber for easier starting. There is one glow plug for each combustion chamber. The glow plugs are located in the cylinder head and extend into the antechamber, figure 10-4. Glow plugs are not used to ignite the fuel directly.

Manufacturers have variations and different nomenclature for various components of the glow plug system. See figure 10-5 for the major components and location of the glow plug system.

How Glow Plug Systems Work

When the ignition key is turned to the run position, the glow plug system is energized, figure 10-6. A light in the dashboard comes on, telling the driver that the glow plugs are heating, figure 10-7. How long the light and glow plugs stay on is determined by how cold the engine is. A temperature sensor mounted in the coolant jacket indicates engine temperature to a control unit. The control unit turns on the glow plug relay, which in turns sends current to the glow plugs. When the control unit has determined that the glow plugs have been on long enough, the indicator light on the dash goes out and current to the glow plugs is turned off. The driver now turns the key to the start position, starting the engine.

The control unit can tell if the engine is running by the increase in electrical circuit voltage due to alternator output. After the engine has started, the glow plug circuit continues to operate—often called the *afterglow* period. The glow plug cycles—turning on and off—to keep the cylinder warm. This aids in providing better combustion; therefore a smoother-running engine when cold. The length of the afterglow period depends on engine temperature.

Should the engine not start, the glow plugs continue to cycle on and off until the battery runs down (approximately 4 to 6 hours).

Types of Glow Plugs

Regardless of type, the glow plugs used on small, high-speed diesels are resistance heaters designed to heat the surrounding air and combustion chamber surface. The glow plug is essentially made up of a heating element, metal sheath, body, and terminal, figure 10-8.

The fast-start glow plug is the most popular type. One type of fast-start glow plug is the 6-volt type. When used in a 12-volt circuit, the glow plug heats very rapidly, which enables the driver to start the vehicle in as little as 6 seconds at 0°F (−18°C), figure 10-9. With this type of glow plug, it is important not to exceed the specified length of time or the glow plug will melt. A major drawback to the fast start (6 volt) glow plug is that it can easily burn out. The latest style incorporates a positive temperature coefficient (PTC) heating coil, figure 10-10. As the temperature of the PTC coil increases, resistance also increases, limiting peak current flow. This helps prevent burn out and reduces the need for glow plug cycling components.

Another fast-start glow plug type uses 12 volts but also uses special resistance materials to provide the needed heat. The wait time at 0°F (−18°C) is about 15 seconds.

The slow-start glow plug is a 12-volt glow plug used in a 12-volt system. It does not have the needed resistance materials or the amperage to provide a fast start. Wait time at 0°F (−18°C) can be as long as 60 seconds. It can be distinguished from the 6-volt glow plug by its smaller terminal connection, figure 10-11.

Types of Controllers

Essentially there are two types of controllers—the solid-state and the electromechanical.

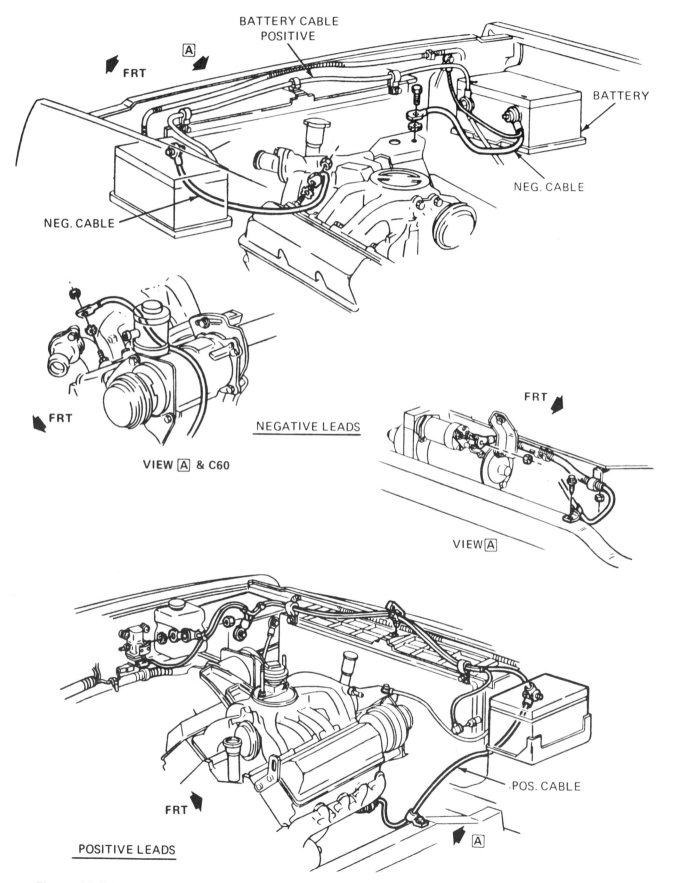

Figure 10-2 Location of battery cables with batteries wired in parallel (*Courtesy of Chevrolet Division*)

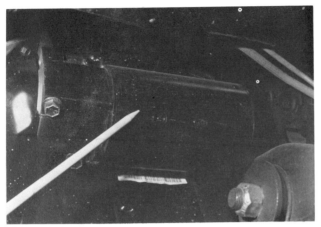

The starter motor shown here is much larger than the comparable gasoline starter motor. This particular motor will draw 600 amperes while cranking at room temperature.

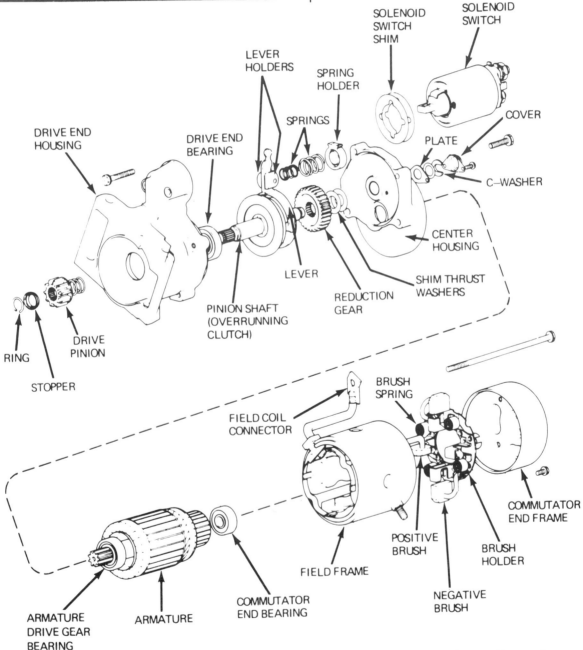

Figure 10-3 Gear reduction starter *(Courtesy of General Motors Product Service Training)*

The solid-state device is a small microprocessor or logic module that senses engine temperature and sends current sent to the glow plugs a specified length of time. It also turns off the wait light at the appropriate time, figure 10-12.

Electromechanical controllers use a series of heating (resistance) elements of various heat outputs to heat up a set of bimetallic switches. These bimetallic switches turn on and off according to the current, hence the heat output of the resistance heaters, figure 10-13. This type of control unit is threaded into an engine coolant passage to sense engine temperature.

Refer to figures 10-14 to 10-17 for typical system operation when the ignition switch is turned to the on position. The following steps explain the operation:

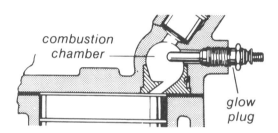

Figure 10-4 Glow plug location (*Courtesy of Volkswagen of America Inc.*)

1. Power flows through the ignition switch to pin 3 of the control switch.

2. The power then flows through resistor R_3, switch S_2, pin 2 (to latching relay), switches S_1, S_3, and pin 6 (to power relay), figure 10-16. While the glow plugs are heating, the resistance heater on switch 1 is also being supplied voltage from the output terminal of the power relay. This starts heating the bimetal spring on switch 1 (this bimetal spring heats and cools very rapidly). After about 10 seconds on a cold engine the bimetal spring is heated enough to cause switch 1 to open, figure 10-17. This breaks the voltage supply to pin 6 and the power relay is deactivated. When there is no voltage feedback from the glow plugs to heat bimetal spring 1, the switch begins to cool and closes again in a few seconds (with a cold engine).

Voltage is again supplied by way of pin 6 to the input of the power relay, which causes the glow plugs to heat again and heat the resistor for switch 1, which opens again when hot enough. This is the cycling process of the glow plugs. The glow plugs help the engine to start quickly, and continue to operate after the engine starts to clear up start-up smoke. This cycling would continue indefinitely were it not for the afterglow signal from the alternator, received after the engine starts. This voltage signal starts heating the bimetal spring on switch 3. After 20 to 90 seconds of receiving voltage, the bimetal spring

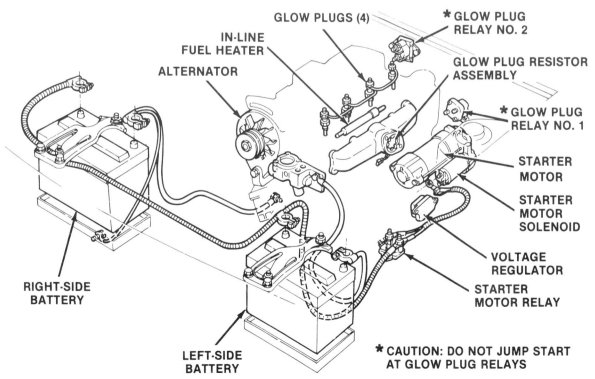

Figure 10-5 Typical glow plug system (note: some components not found on all systems) (*Courtesy of Ford Motor Company*)

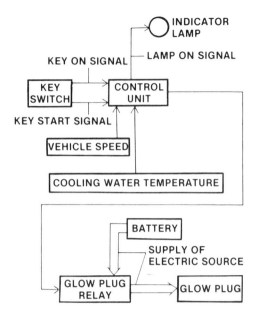

Figure 10-6 *Courtesy of Ford Motor Company*

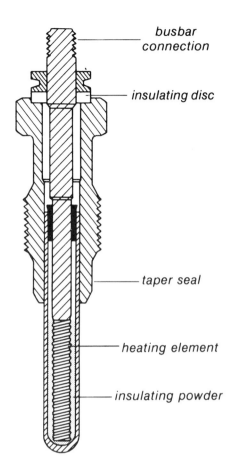

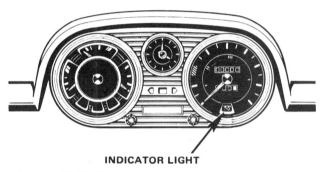

Figure 10-7 Glow plug indicator light (*Courtesy of Mercedes-Benz of North America Inc.*)

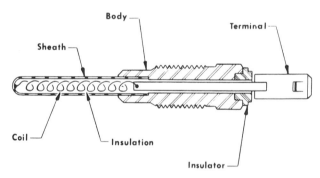

Figure 10-8 Glow plug components. (A—*Courtesy of Volkswagen of America Inc.*; B—*Courtesy of General Motors Product Service Training*)

A glow plug relay is often used to carry the large amount of current needed by the glow plugs.

is heated enough to cause switch 3 to open and shut the glow plug system off.

If the vehicle did not start, glow plugs would continue to cycle on and off until the batteries died. To correct this condition, an after-glow timer is added (see figure 10-17). When the ingition switch is turned to run, a voltage is applied to the after-glow timer. The after-glow switch opens after approximately 2 minutes, depending on engine temperature. This prevents the glow plugs from turning on.

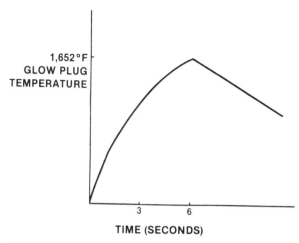

Figure 10-9 Fast-start glow plug temperature vs. time (*Courtesy of Ford Motor Company*)

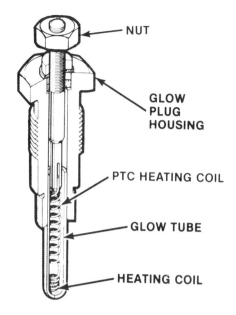

Figure 10-10 Fast-start glow plug with PTC heating coil (*Courtesy of Ford Motor Company*)

CAUTION: To reset the after-glow timer, the ignition switch must be turned off and the after-glow timer allowed to cool (about 1 minute). Failure to do this will result in hard starting.

Solid-State Glow Plug Operation

A typical glow plug system with solid-state controller consists of glow plugs, the control module, two relays, a glow plug resistor assembly, figure 10-18, coolant temperature switch, clutch, neutral switches, and connecting wiring, figure 10-19. Relay power and feedback circuits are protected by fuss links in the wiring harness. The control module is protected by a separate 10A fuse in the fuse panel.

When the ignition switch is turned to the on position, a wait-to-start lamp in the instrument cluster lights. When the lamp lights, relay 1 also closes and full system voltage is applied to the glow plugs. If engine coolant temperature is below 30°C (86°F), relay 2 also closes. After 3 seconds, the control module turns off the wait-to-start lamp, indicating that the engine is ready for starting. If the ignition switch is left in the on position about 3 seconds more without cranking, the control opens relay 1 and current to the plugs stops to prevent overheating. However, if coolant temperature is below 30°C (86°F) when relay 1 opens, relay 2 remains closed to apply reduced voltage

The fast start diesel glow plug control system uses 6 volt glow plugs with controlled pulsing current applied to them for starting. The slow start system used steady current applied to 12 volt glow plugs. In either case the correct glow plug should be used for proper starting. The illustration shows the glow plug identification.

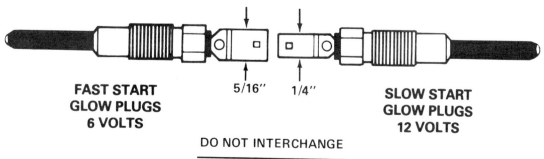

Figure 10-11 Diesel engine glow plug identification (*Courtesy of General Motors Product Service Training*)

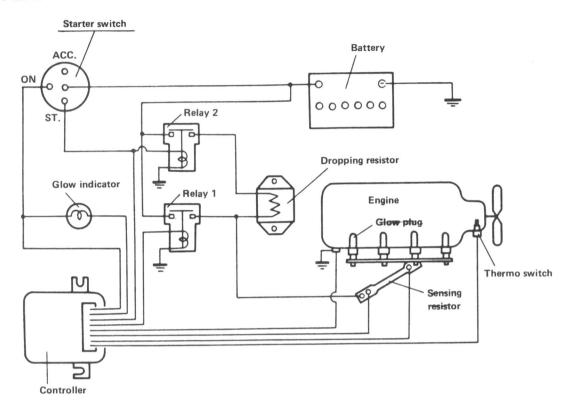

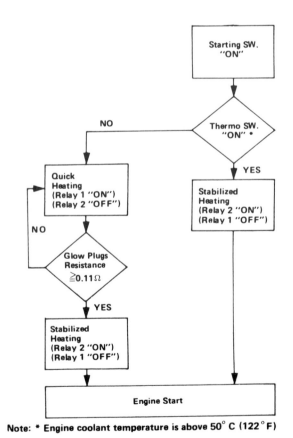

Figure 10-12 Glow plug circuit controlled by solid-state controller (*Courtesy of General Motors Product Service Training*)

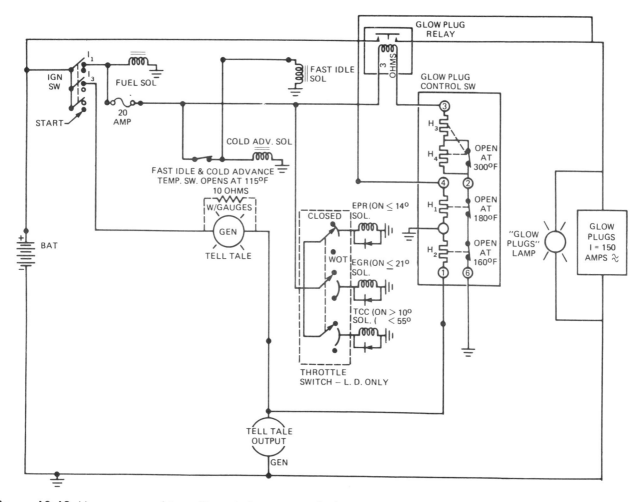

Figure 10-13 Heaters open bimetallic switches at specified temperatures in glow plug controller. *(Courtesy of General Motors Product Service Training)*

to the plugs through the glow plug resistor, until the ignition switch is turned off.

CAUTION: Leaving the ignition switch on starting will run the batteries down.

When the engine is cranked, the control module cycles relay 1 intermittently. Thus, glow plug voltage will alternate between 12 and 4 volts, during cranking, with relay 2 closed, or between 12 and 0 volts with relay 2 open. After the engine starts, alternator output signals the control module to stop the no. 1 relay cycling and the afterglow function takes over.

If the engine coolant temperature is below 30°C (86°F), the no. 2 relay remains closed. This applies reduced voltage (4.2 to 5.3) to the glow plugs through the glow plug resistor. When the vehicle is underway (clutch and neutral switches closed), or coolant temperature is above 30°C (86°F), the control module opens relay 2, cutting off all current to the glow plugs.

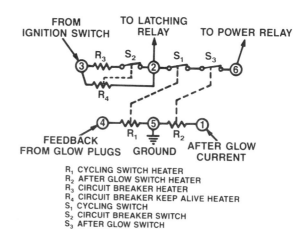

R_1 CYCLING SWITCH HEATER
R_2 AFTER GLOW SWITCH HEATER
R_3 CIRCUIT BREAKER HEATER
R_4 CIRCUIT BREAKER KEEP ALIVE HEATER
S_1 CYCLING SWITCH
S_2 CIRCUIT BREAKER SWITCH
S_3 AFTER GLOW SWITCH

Figure 10-14 *Courtesy of Ford Motor Company*

Temperature Sensor

The engine temperature sensing unit is the thermistor. A thermistor is a resistor that changes its resistance to current flow according to engine temperature. The con-

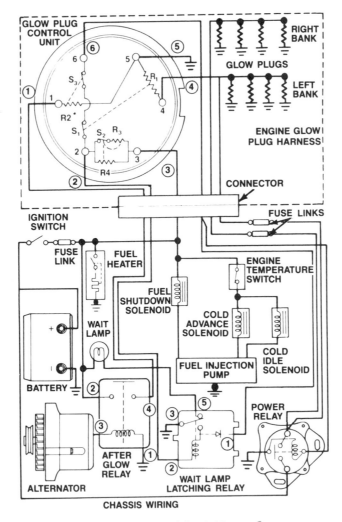

Figure 10-15 *Courtesy of Ford Motor Company*

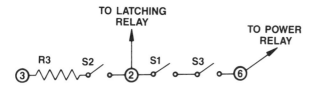

Figure 10-16 *Courtesy of Ford Motor Company*

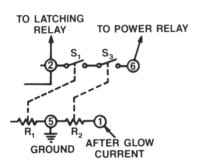

Figure 10-17 *Courtesy of Ford Motor Company*

troller (solid-state) senses the change in temperature through resistance and can determine how long the glow plugs should remain on.

BLOCK HEATER

Block heaters are usually part of cold weather packages that aid starting in cold climates. They are found in place of one of the block's core plugs with a resistance element immersed in the coolant, figure 10-20. The block heater is powered by household current.

ENGINE-CRANKING SYSTEM SERVICE

Service and testing of the diesel engine-cranking system is nearly identical to the same procedures used on gasoline engines. This section only mentions the differences.

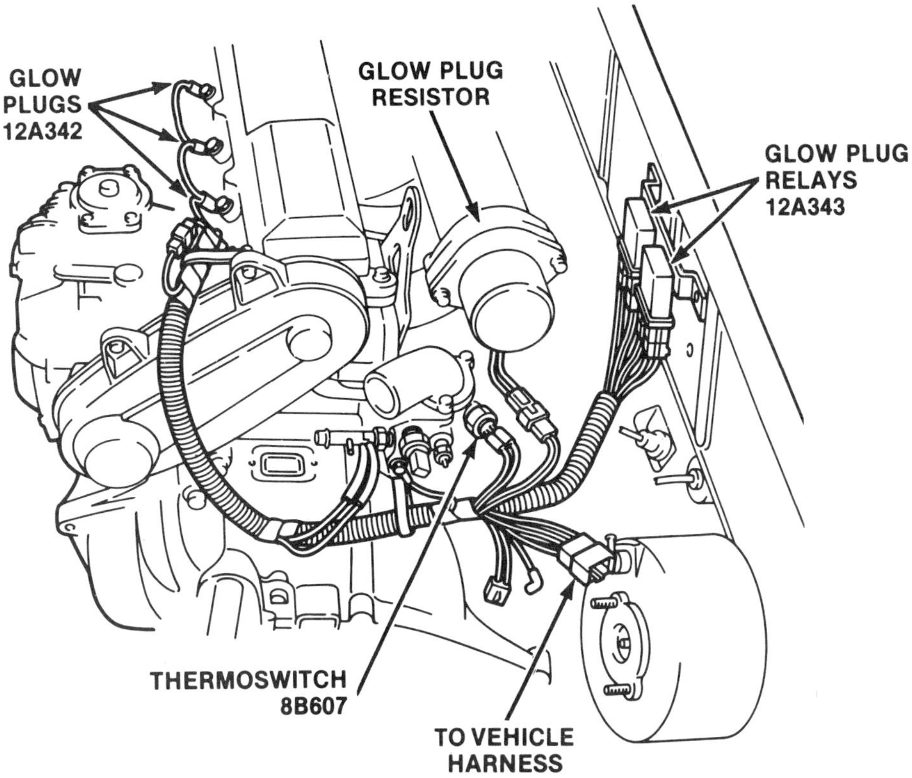

Figure 10-18 Glow plug system component location (*Courtesy of Ford Motor Company*)

CAUTION: Starting fluids must never be used on the diesel engines used in cars and small trucks. Because of the high CR and hot glow plugs, an explosion may result, causing personal injury and engine damage.

Jump Starting

When jump starting a diesel-powered vehicle, follow all the precautions you would use on a gasoline-powered vehicle and in addition:

- Be certain the jumper cables can carry the extra current needed.
- If two batteries are used on the vehicle, connect the positive jumper cable to the positive battery terminal closest to the starter. Connect the negative cable to a good ground other than the negative post. This reduces resistance losses.
- Never use 24 volts to boost the starter system. This can damage the glow plug controller.

Cranking Speed

Cranking speed is extremely important for a diesel engine to start, either cold or hot. A tachometer is generally

used to determine cranking speed. Simply disconnect the fuel shut-off wire or use the manual stop to prevent fuel flow. With the tachometer connected, crank the engine and note engine RPM. At low cranking speeds the tachometer may not be accurate. If in doubt, follow this procedure. It requires the following:

- a compression gauge
- a watch
- an assistant
- appropriate hand tools

Test the tachometer accuracy by following these procedures:

1. Install the compression gauge in any cylinder.

2. Disconnect the fuel shut-off wire or use the manual stop, if equipped, to prevent fuel flow.

3. Depress the pressure relief valve on the compression gauge.

4. Have an assistant crank the engine for 3 seconds. Then, without stopping, count the number of puffs at the compression gauge that occur in the next 10 seconds. Multiply the number of puffs in the 10-second period by 12. That number is the cranking RPM.

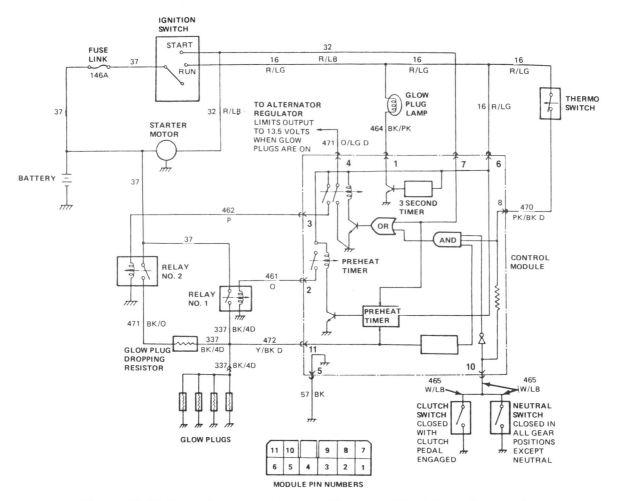

Figure 10-19 Glow plug wiring schematic *(Courtesy of Ford Motor Company)*

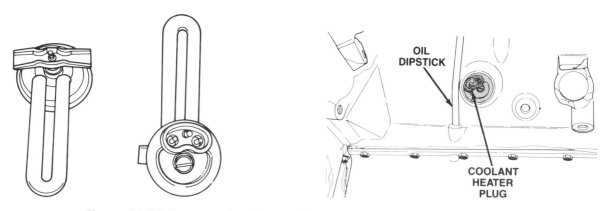

Figure 10-20 Engine coolant heater *(Courtesy of Ford Motor Company)*

Battery Capacity and Starter Draw Tests

Follow the manufacturer's procedures for connecting a volt-amp meter to the starting system. Be certain the volt-amp meter has the capacity to handle the high-amperage current found on larger diesel engines. The battery capacity can be as high as 1,000 cold cranking amps, and starter load 400 amps, under normal conditions. If two batteries are used, they must be disconnected before doing a battery load test. The troubleshooting of this system is the same as with gasoline-powered vehicles, but the specifications are higher.

GLOW PLUG SYSTEM DIAGNOSIS

Each vehicle manufacturer has its own glow plug system, and therefore its own diagnostic procedures. What follows is general procedure for diagnosis of this system. For further information, consult the manufacturer's service manual.

A glow plug system problem is likely when the owner complains of hard starting or the glow plug wait light remains on. If the problem is hard starting, be certain the correct starting procedure is being followed.

Next, it must be determined whether the fuel system or the glow plug system is at fault. Generally, when the glow plugs are turned on, the glow plug relay makes a noticeable click. If no noise is heard, chances are the glow plug system is not operating. If the fuel system is suspect, consult Chapter 18, the service manual, or both.

A test light can be used to determine if the glow plugs are working. With the test light grounded, insert the test light probe into a glow plug wire to check for continuity.

If the light lights, check the other cylinders in the same manner.

If the test light failed to light, check for continuity at the glow plug relay. If no power is present, check the control circuit of the glow plug relay. If the control circuit is working, the relay may be defective. Be aware that manufacturers may have the system fused and that fusible links may be located throughout the system, figure 10-21.

If the glow plug circuit is working, the glow plugs themselves may be defective. With an ohmmeter, connect one lead to the terminal and the other lead to the body. Read the resistance and compare to specifications. A visual inspection may also be necessary to check for excessive carbon buildup. A glow plug that has excessive carbon deposits on its tip can give the correct ohmmeter reading but still not work. Clean the tip with cleaner that removes carbon.

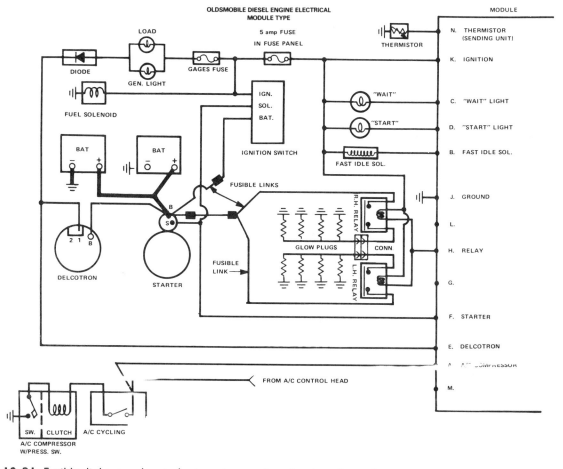

Figure 10-21 Fusible links are located at various points in the glow plug system (*Courtesy of General Motors Product Service Training*)

CAUTION: Do not wire brush the tip.

When installing glow plugs, torque them in to specifications and use an antiseize compound on the threads. Remember that they may be in for at least 50,000 miles.

Defective controllers can be checked with an ohmmeter, jumper leads, and the process of elimination. Carefully follow the manufacturer's procedure for diagnosing faulty controllers.

When replacing a glow plug temperature sensor or an electromechanical controller that threads into a coolant jacket, always let these devices lie on the intake manifold for approximately 10 minutes. This allows the controller or thermal switch to adjust to engine temperature, reducing the chances of coolant leakage and glow plug failure. Remember, some systems use 12 volts applied to a 6-volt glow plug. If the glow plugs stay on longer than necessary, they will be damaged.

With the inductive pickup over the glow plug relay and the meter set to position 1, turn the ignition switch to the run position. The ammeter will indicate the current draw for the glow plug system.

An inductive ammeter can be used to check overall glow plug system performance. Clip the inductive lead over the wire leading the glow plug relay to the glow plugs. Turn the ignition switch to the run position and observe the ammeter. A typical reading for 4 cylinders is 35–40 amps; for 8 cylinders with 6-volt glow plugs, 125–130 amps.

BLOCK HEATER SERVICE

Check all block heater connections. With an ohmmeter, check the continuity of the resistance element, figure 10-22.

SUMMARY

Diesel engine starting efficiency decreases in cold weather. Anything that takes heat away from the combustion chamber reduces the chances of the air/fuel mixture igniting.

Manufacturers incorporate starting aids into the engine for cold weather starting. In addition to a high CR, a heavy-duty starter system, glow plug system, and engine block heater are used.

The starter circuit is identical to a gasoline starter circuit. Variations may include extra battery, larger cables, and starter motor.

The glow plug system adds heat to the combustion chamber before engine cranking and provides smooth engine operation when cold.

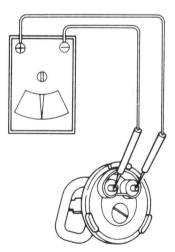

Figure 10-22 Using an ohmmeter to check the resistance element (*Courtesy of General Motors Product Service Training*)

CHAPTER 10 QUESTIONS

1. List three differences between starting a gasoline engine and a diesel engine.
2. Name the major components of the glow plug system.
3. Describe how the glow plug system works.
4. Why are ether-type starting aids not to be used?
5. What is the difference between a fast glow plug system and a slow glow plug system?
6. What cautions must be observed for jump starting a vehicle?
7. Why is antiseize compound used on glow plug threads?

Objectives

In this chapter you will learn:

- Why HC, CO, and NOx emissions are low on a diesel engine
- To identify diesel engine emission components
- The operation of the crankcase ventilation system
- The operation of the EGR valve
- Three locations for a vacuum pump
- How to inspect and service the crankcase ventilation system, the EGR system, and the vacuum pump

DIESEL ENGINE EMISSIONS

As discussed in Chapter 1, the diesel produces lower amounts of hydrocarbons (HC), carbon monoxide (CO), and oxides of nitrogen (NOx) than the comparable gasoline engine. HC and CO are lower because of more complete combustion of the fuel and air. HC and CO are further reduced by routing the blow-by gases in the crankcase to the intake manifold where they can be reburned.

NOx is lower because the peak temperature is not maintained very long. However, because the government measures emissions by volume (grams per mile), larger-displacement diesels use an exhaust gas recirculation (EGR) system to lower the emissions.

Smoke, or particulate emissions, occurs when there is insufficient air to completely burn the fuel. The leftover fuel is heated and changed to soot. Anything that lowers

With the HC and CO meter set on the LO scale, the HC and CO content is extremely low.

the amount of air taken into the cylinder, increases the ratio of fuel to air, and changes injection timing, increases particulate output. When government legislation requires the particulate level to be lowered, the larger-displacement diesels suffer because they burn larger quantities of air and fuel, producing more soot.

There are several factors that the engine designer varies to provide low emission levels with high performance and good fuel economy. Some of these factors are the shape of the combustion chamber, the location and angle of the nozzle, the injection rate and nozzle spray pattern, injection timing, and camshaft timing.

Generally, with the diesel engine, emission levels have been lowered through engine refinements and not through the addition of extra systems. This has also had a positive effect on economy and performance.

Of the factors previously listed, the technician has control over and the responsibility to see that injection timing and nozzle spray pattern are correct. However, there are several other factors that affect emission levels and engine performance that the technician must inspect and correct. These may be as simple as changing a dirty air cleaner, or as complicated as changing piston rings to correct the problem.

CRANKCASE VENTILATION SYSTEM

This system performs the same task as the positive crankcase ventilation (PCV) system on a gasoline engine. Blow-by gases from the crankcase are drawn into the intake manifold where they mix with the incoming air, figures 11-1, 11-2. Drawing the blow-by gases out of the crankcase also reduces pressure buildup, which forces oil through the seals. Instead of a PCV valve, some diesel engines use a device called a crankcase depression regulator (CDR), figure 11-3.

Fresh air is drawn through a filter into the crankcase, where the fresh air and blow-by gases mix, and then up into the valve cover through the CDR, into the intake manifold.

The CDR is designed to maintain a specified pressure in the crankcase. Too little vacuum (too much pressure) in the crankcase causes oil leaks; too much vacuum and oil is drawn into the air intake passage.

At idle, the CDR is fully open to allow blow-by gases to enter the intake manifold. Since there is very little intake vacuum at idle speed, the CDR must be fully open to prevent pressure buildup in the crankcase.

As engine speed increases, intake vacuum increases, pulling a spring-loaded diaphragm closer to the outlet tube which prevents too much vacuum in the crankcase and the possible suction of oil into the intake passages, figure 11-3. Many manufacturers use a vapor separator to

The CDR directs the blow-by gases from the oil filler tube to the intake manifold.

prevent loss of oil through the combustion chamber. The diesel can run on its own crankcase oil, preventing the engine from shutting off. This can be a hazardous situation.

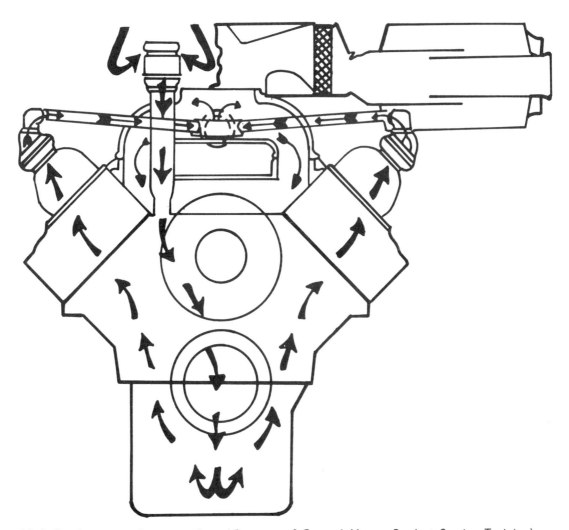

Figure 11-1 Crankcase ventilation gas flow (*Courtesy of General Motors Product Service Training*)

Always be sure the vapor separator is connected and working properly.

EXHAUST GAS RECIRCULATION SYSTEM

The EGR system is designed to reduce NOx (oxides of nitrogen) emissions. As the name implies, some of the exhaust gases are routed back to, and mixed with, intake air. Diluting the intake air reduces peak combustion temperature. The oxygen content of the exhaust gases is very low and will not contribute anything to the combustion process. When combustion is started, the temperature rises rapidly, causing the gases to expand. The spent exhaust gases cannot burn, but the increase in temperature causes

them to expand. As the spent gases expand, they absorb the heat of combustion, lowering peak combustion temperature (approximately 500°F [260°C]).

The EGR valve is generally located on the air crossover assembly or intake manifold. From this vantage point, exhaust gases are routed into the intake air stream, figure 11-4.

At idle, the EGR valve is fully open. It is pulled and held open by vacuum supplied by the vacuum pump, and regulated by the vacuum regulator valve (VRV). The VRV allows full vacuum to pass through with the throttle shaft at the idle position, and closes off vacuum at full throttle position. At intermediate throttle positions, the valve position (and therefore the vacuum signal) is proportional to throttle position. As throttle position moves to maximum fuel, the strength of the vacuum signal decreases and the EGR valve gradually closes, figure 11-5.

Working in conjunction with the EGR system on some engines is the exhaust pressure regulator (EPR) system. The EPR valve is located in the exhaust system, figure 11-6. It is designed to restrict exhaust gas flow and increase back pressure at certain throttle positions. At idle, the strong vacuum signal sent by the VRV closes the EPR valve. As the throttle position increases, the EPR also opens, reducing back pressure.

With the EPR closed at idle, more of the exhaust gases are forced to go through the EGR valve. Remember that at idle there is very little intake vacuum, too little to draw the necessary amount of exhaust gases. So at idle, the EPR is fully closed and the EGR is fully opened. As the throttle position increases, the EPR gradually opens and the EGR gradually closes. There may be other valves within the two systems to smooth and modify the operation when different conditions are encountered.

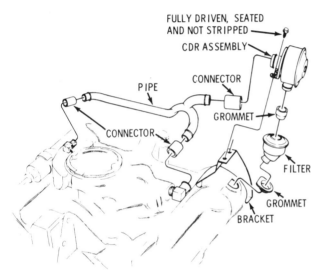

Figure 11-2 Crankcase ventilation components (Courtesy of General Motors Product Service Training)

CRANKCASE DEPRESSION REGULATOR VALVE
V-TYPE DIESEL ENGINE

1 INLET PORT (2) (GASES FROM) CRANKCASE

2 MOUNTING BRACKET

3 COVER DIAPHRAGM

4 BODY

5 SPRING

6 OUTLET TUBE (GASES TO INTAKE) MANIFOLD

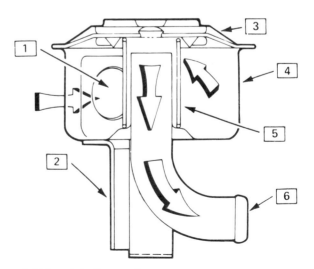

Figure 11-3 CRR valve (Courtesy of Oldsmobile Division)

VACUUM PUMPS

Many control and assist devices on the car are vacuum powered, such as the emission control system or power-

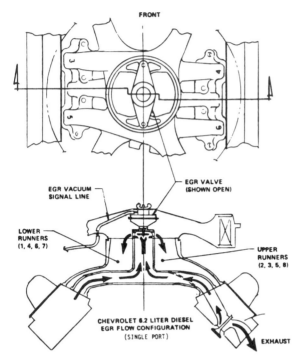

Figure II-4 EGR valve and exhaust gas flow (*Courtesy of General Motors Product Service Training*)

assisted brakes. To operate these devices on the minimal amount of vacuum that the diesel produces is impractical; so a vacuum pump is used to provide the needed vacuum. Two different types of vacuum pumps are used—the diaphragm and the vane.

The diaphragm type may be driven by the engine or electrically driven. When driven by the engine, the vacuum pump runs off the camshaft or accessory belt. When driven by the camshaft, figure 11-7, the vacuum pump also drives the oil pump.

CAUTION: Never run the engine with the vacuum pump removed.

Unlike the gasoline engine with the distributor removed, the diesel will run—until it seizes.

The other typical location for the diaphragm vacuum pump is at the front of the engine. Here it is belt driven.

The diaphragm is driven by an eccentric that forces it to move back and forth. This causes air to flow into the inlet tube, through the pump, and to exhaust out the rear port.

The electrically operated vacuum pump has the advantage of producing the required amount of vacuum when needed, regardless of engine RPM, and can be mounted where space permits, figure 11-8.

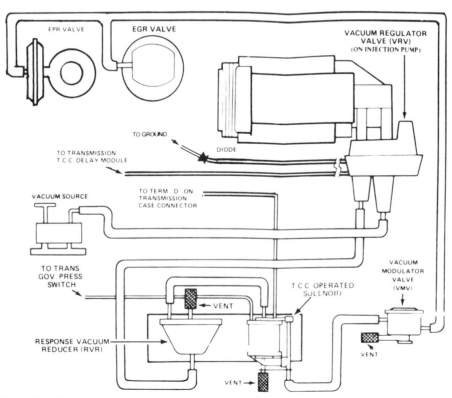

Figure II-5 EGR-EPR emission system (*Courtesy of General Motors Product Service Training*)

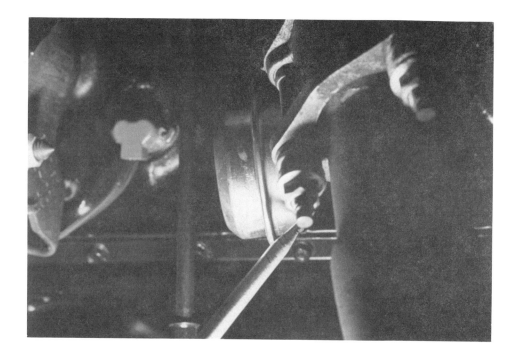

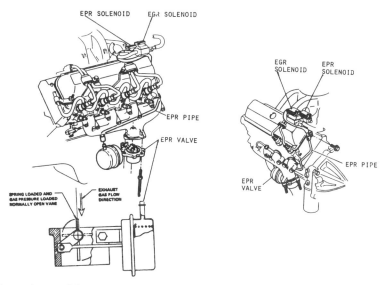

Figure 11-6 The EPR valve is located between the exhaust manifold and exhaust pipe. (*Drawing Courtesy of General Motors Product Service Training*)

The vane type is mounted on the back of the alternator and is driven by the rotor shaft. Vanes draw air through an inlet port, then out to exhaust. Engine oil is used to lubricate the vanes.

CRANKCASE VENTILATION SYSTEM SERVICE

Periodically, the crankcase ventilation system must be serviced. This involves cleaning or replacing the breather filters and the CDR (PCV) valve. Follow the manufacturer's scheduled service for maintenance of these items.

When this system does fail to operate, a common complaint is excessive engine oil leakage (too little vacuum) or oil in the air intake passage (too much vacuum). To check the crankcase ventilation system requires the following:

- water manometer
- tachometer
- manufacturer's service manual

Service the crankcase ventilation system by following these procedures:

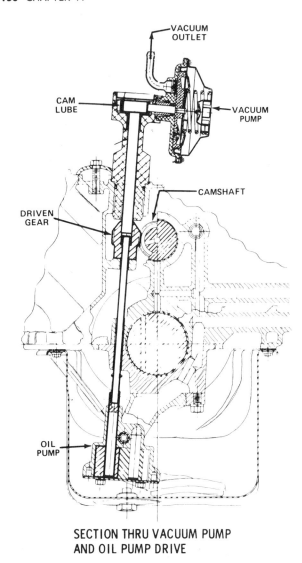

SECTION THRU VACUUM PUMP
AND OIL PUMP DRIVE

Figure 11-7 Camshaft-driven diaphragm vacuum pump
(Courtesy of General Motors Product Service Training)

1. Remove oil dipstick and connect water manometer over the dipstick tube.
2. Set engine to specified RPM.
3. Read water manometer and compare to specifications. To read the manometer, add the amount the water traveled in the left column to the amount the water traveled in the right column to obtain the total, figure 11-9.

A typical range would be 3 inches of water (7.62cm) maximum, to 1 inch of water (2.54cm) minimum.

EGR/EPR SYSTEM SERVICE

According to scheduled service, the EGR/EPR must be cleaned and checked. This involves cleaning carbon deposits off the valves, passages, and checking valve operation. The latter requires the following:

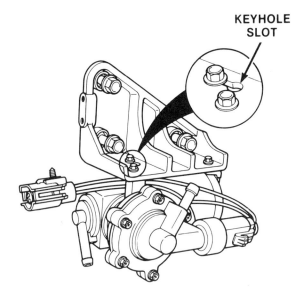

Figure 11-8 Electrically driven diaphragm vacuum pump
(Courtesy of Ford Motor Company)

- vacuum pump
- manufacturer's service manual

Clean the EGR/EPR system by following these procedures:

1. Connect the vacuum pump to the valve and apply the specified vacuum.
2. Observe valve movement.

When either of the EGR/EPR valves do not work, common complaints are loss of power and heavy smoke, particularly on acceleration. The following lists will aid in diagnosis.

EGR valve will not open.

1. Binding or stuck EGR. Clean or replace EGR.
2. No vacuum to EGR. Check operation of VRV and other control valves. Inspect vacuum lines and vacuum pump for proper operation.

EGR valve will not close or EPR will not open. (Heavy smoke on acceleration.)

1. Binding or stuck EGR or EPR valve(s). Clean or replace EGR/EPR valve.
2. Constant high vacuum to EGR/EPR valves. Check VRV and any other valves within the system.

EGR valve opens partially.

1. Binding EGR valve. Clean or replace EGR valve.
2. Low vacuum to EGR. Check VRV and any other valves within the system. Check vacuum and connecting hoses.

Loss of power and heavy smoke on acceleration.

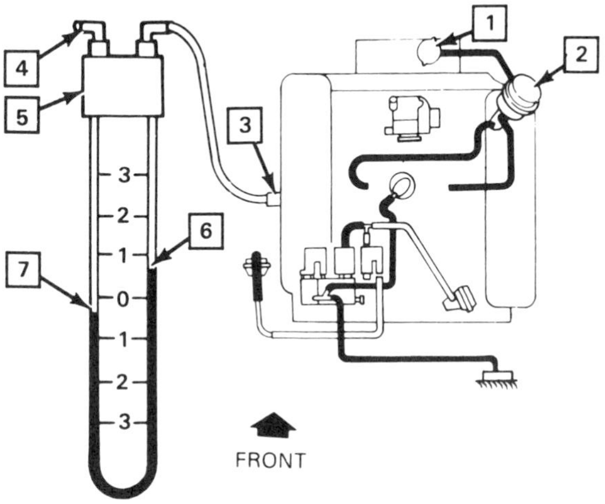

1	OIL FILL PIPE
2	CRANKCASE DEPRESSION REGULATOR VALVE
3	ENGINE OIL DIPSTICK TUBE
4	VENT TO ATMOSPHERE
5	MANOMETER - J23951
6	½" ABOVE ZERO (SEE EXAMPLE)
7	½" BELOW ZERO (SEE EXAMPLE)

Figure 11-9 One end of the water manometer is connected to the oil dipstick opening, the other end is open. To determine the crankcase pressure, add both columns together (½" + ½") to get the total (1" of vacuum). *(Courtesy of Chevrolet Division)*

1. Binding or stuck EPR valve. Clean or replace EPR.
2. Constant high vacuum to EPR. Check VRV and any other valves within the system. Check vacuum hose routing.

VACUUM PUMP SERVICE

Service on vacuum pumps is limited because they are generally not rebuildable. When defective, they are replaced. Checking a vacuum pump requires the following:

● vacuum gauge
● manufacturer's service manual

Check the vacuum pump by following these procedures:

1. Visually inspect all connections first. If vacuum pump is belt driven, be sure the drive belt has the proper tension.
2. Connect the vacuum gauge to the vacuum pump and run the engine. Compare reading to specifications (approximately 20 in. [50.8cm] of mercury).

SUMMARY

HC and CO emissions are so low on a diesel engine that no add-on devices are needed to control these pollutants. On larger diesels, the use of an EGR valve may be required to reduce NOx emissions.

A crankcase ventilation system is used to prevent excessive pressure buildup in the crankcase and to burn blow-by gases. A crankcase depression regulator may be used in place of a PCV valve.

Vacuum pumps are used on diesel engines to operate accessories and emission control devices. The two main types are the vane and diaphragm. The diaphragm can be electrically driven or engine driven. The vane type mounts on the back of the alternator and is driven by the rotor shaft.

CHAPTER 11 QUESTIONS

1. Explain why HC, CO, and NOx emissions are low on a diesel engine.
2. Identify diesel engine emission components.
3. Explain the operation of the crankcase ventilation system.
4. Explain the operation of the EGR valve.
5. Explain the operation of the EPR valve.
6. Name three locations for a vacuum pump.
7. Why must an engine with a camshaft-driven vacuum pump never be run with the pump removed?

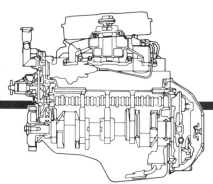

chapter 12
Basic Engine Tests and Disassembly

Objectives

In this chapter you will learn:
- **To identify basic engine configurations**
- **To perform engine visual checks**
- **To perform a compression test**
- **To successfully remove a diesel engine from the chassis**
- **To successfully disassemble a diesel engine**

BASIC ENGINE CONFIGURATIONS

Many of the service procedures used on gasoline engines are the same for diesel engines, particularly in the area of engine overhaul. However, there are cautions and procedures unique to the diesel that must be followed or the results may be disastrous.

Diesel engine configurations for cars and small trucks are classified according to the shape of the engine block, number of cylinders, and the location of the camshaft. Like the gasoline engine, the engine block may be V-shaped or straight in-line configuration. The V-shaped block has six or eight cylinders and the camshaft is placed within the block, figure 12-1. This design is used in the larger-displacement engines. The straight in-line configuration is either a four or five cylinder and usually uses an overhead camshaft design, figure 12-2. Some in-line

engines do have the camshaft within the block. The in-line design is used for small-displacement engines.

ENGINE VISUAL CHECKS

Before a wrench or socket is used on an engine, it is important to observe the engine and related components. Visually checking the engine before any repairs are done may give valuable clues to the cause of the problem. Be sure the problem is within the engine to prevent needless repairs.

Before draining, check and inspect the oil level for the proper amount and possible contamination. If it is low, look for signs of leakage or burning. If you have access to the service history of the vehicle, find out when the oil was checked, changed, and if there has been a change

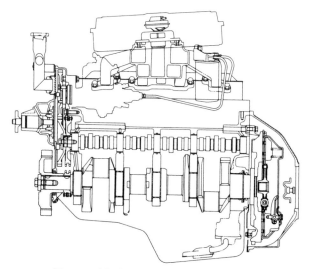

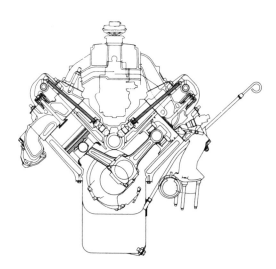

Figure 12-1 V-shaped engine block *(Courtesy of General Motors Product Service Training)*

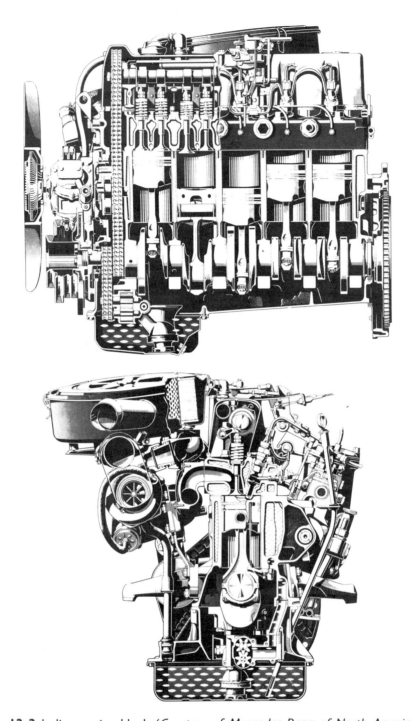

Figure 12-2 In-line engine block *(Courtesy of Mercedes-Benz of North America Inc.)*

in consumption of oil or fuel. Contamination of the oil may result from the mixing of oil with antifreeze or diesel fuel. Possible sources for this condition may be a leaking head gasket, oil cooler, or injection pump.

Check and inspect the engine coolant level for the proper amount and possible contamination. Inspect the radiator and heater hoses, the radiator and radiator cap for signs of deterioration and leakage. Observe the condition of the coolant to be sure it is set for the proper temperature and is not contaminated.

Observe the condition of the fuel system and note any signs of leakage. Inspect the lines for any kinks or restrictions.

Finally, note the position of brackets, wires, hoses, and so on. If in doubt, mark or write down the position of the components before disassembly. For example, use masking tape to number wires and hoses. The important point of all this is to observe and obtain clues to the cause of the problem before they are lost.

COMPRESSION TEST

One of the quick, accurate tests of engine condition is the compression test. It is very similar to the compression test performed on gasoline engines. A compression test is performed when in doubt of engine condition or to verify engine cylinder compression.

This task requires the following:

- appropriate hand tools
- a compression gauge designed for diesel use with adapters
- air cleaner guard
- manufacturer's service manual

CAUTION: Do not use a compression gauge designed for gasoline engines. The maximum pressure of these gauges is lower (about 275 psi, [1895 kPa]) than that of a diesel compression gauge (800 psi, [5516 kPa]). Serious injury may result from improper use.

Perform the compression test by following these procedures:

1. Be sure the vehicle is at operating temperature.
2. Be sure battery(ies) are in good condition and fully charged.
3. Remove the air cleaner and install the air cleaner guard.

CAUTION: Always install a guard over the air intake passage. The guard is needed to avoid personal injury and prevent debris from entering the intake manifold.

---

4. Disconnect the wire to the fuel shut-off on the injection pump.
5. Manufacturers require that cylinder compression be measured either through the glow plug hole or nozzle cavity. Therefore, either disconnect the glow plug wires and carefully remove all glow plugs, or remove all injection nozzles, nozzle gaskets, and cap all exposed fuel openings.
6. Install the proper adapter and securely attach it to the compression gauge, figure 12-3.

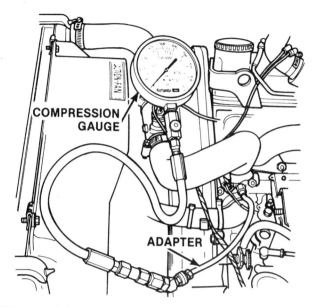

Figure 12-3 Attach adapter to compression gauge. *(Courtesy of Ford Motor Company)*

7. Crank the engine and observe the compression gauge needle for at least six compression strokes. Record the reading and repeat the procedure on the remaining cylinders:

cyl #1 _____   cyl #5 _____
cyl #2 _____   cyl #6 _____
cyl #3 _____   cyl #7 _____
cyl #4 _____   cyl #8 _____
man. spec. _____

8. Compare readings to manufacturer's specifications.

Under normal conditions, cylinder pressure builds evenly and rapidly. A low compression reading usually indicates worn piston rings, poor valve seating, head gasket failure, or valve train failure. Normally on a gasoline engine, a wet compression test (adding oil to the cylinder, then taking a compression reading) would be done to pinpoint the cause.

CAUTION: On a diesel this may be disastrous!

Because the compression ratio is so high, adding oil can cause a hydraulic lock. Never put oil into the cylinder; it may lead to internal damage to the engine. The alternative to the wet compression test is to observe the compression gauge as the engine is being cranked. If the pressure remains low after the first compression stroke, a leaking valve may be the cause. If pressure builds slowly after the first stroke but does not reach specifications, the piston rings may be the cause.

When a low compression reading is indicated, it usually means that major engine repairs are required.

9. If everything is normal, install the glow plugs to the proper torque, if they have been removed. If you have removed the nozzle, always install a new sealing gasket first and then properly install the nozzle.

10. Reconnect the fuel shut-off wire and install the air cleaner.

11. Check and operate the vehicle to ensure customer satisfaction.

ENGINE REMOVAL

Removing the engine and related engine components requires the following:

● appropriate tools
● appropriate support and lifting equipment
● manufacturer's service manual

CAUTION: If you must clean the engine, be sure the engine is cold, then apply degreaser and water. Do not direct spray on the injection pump. Damage to the injection pump will result if it is suddenly cooled or heated.

Remove the engine by following these procedures:

1. With fender covers in place, disconnect the negative battery cable(s) and engine ground strap.

2. Mark the hood and remove it from the hinges.

3. Remove the air cleaner and install the air cleaner guard.

4. Drain the cooling system and engine crankcase.

5. Disconnect the radiator hoses, engine oil, and transmission oil lines; remove radiator supports and radiator.

6. Disconnect the heater hoses, vacuum lines, power steering hoses, fuel hoses, and air conditioner compressor.

7. Disconnect the engine electrical harness and starter cables.

8. Disconnect the throttle linkage.

9. Disconnect the exhaust pipes.

10. Support the transmission and remove the transmission bolts.

11. Make sure that whatever support and lifting equipment you have can withstand the extra weight of the diesel engine. Secure the lifting device at the appropriate points on the engine, figure 12-4.

12. Disconnect the engine mounts and carefully lift engine out of vehicle.

13. Mount the engine on the stand.

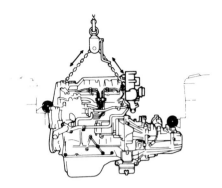

Figure 12-4 Be certain the engine is properly secured before lifting.

ENGINE DISASSEMBLY

This is a general guide for disassembling an engine and is meant to supplement the manufacturer's procedures. Throughout this guide, there are cautions that should be followed to prevent personal injury and damage to the motor.

1. Remove the fuel filter and fuel lines. Mark the position of the fuel injection pump and remove it. Do not bend or kink the injection lines, and cap all openings to prevent contamination.

2. Remove the intake manifold. Note the position of any brackets and harnesses attached to it.

3. Remove the injection nozzles and note which cylinder they came from. Cap all openings to prevent contamination.

4. Remove valve cover(s), rocker arms, and pushrods.

CAUTION: Indicate which side is the top of the pushrod, since some manufacturers harden only one side.

---

5. Remove hydraulic lifters and their holders.

CAUTION: Be sure to note which bore each lifter came from, since some lifters may be oversize.

---

6. Remove the water pump.

7. Remove the crankshaft pulley and harmonic balancer. Use a puller designed for removing harmonic balancers.

8. Remove the timing cover. Note the position of the timing gears and locate the timing marks.

9. Remove timing gears, and chain or belt.

10. Remove the starter and flywheel.

CAUTION: The flywheel on a diesel engine is heavier than the comparable gasoline engine flywheel.

Be sure you have help and a good grip on the flywheel.

11. Remove the oil cooler and oil dipstick tube.
12. Loosen each cylinder head bolt in the proper sequence one turn at a time until all bolts are loose. Remove the cylinder heads.

CAUTION: The cylinder head on the diesel engine is heavier than the comparable gasoline engine cylinder head. Be sure you have help and a good grip on the cylinder head.

13. Clean off the carbon ridge inside the cylinder bore. A good carbon cleaner (such as carburetor cleaner) will soften the carbon ridge. Scrape off the remaining carbon with a steel brush. If there is still a ridge, carefully use a ridge reamer. Clean the cylinder bore.
14. Remove the oil pan.
15. Note the position of the rear main oil seal cover, then remove it.
16. Mark the connecting rod caps by cylinder number, then remove. Cover each connecting rod stud with pastic or rubber tubing, figure 12-5.
17. Mark the position and direction of the main bearing caps, then remove them.
18. Remove the crankshaft.
19. Remove the pistons and be sure to note which cylinder bore each piston came from.
20. Remove the soft plugs and block heater.
21. Remove the oil spray tubes.

SUMMARY

Diesel engines for automobiles use either the V or inline configuration. The position of the camshaft may be in the block or above the cylinder head.

Before disassembling any component (particularly the engine), the engine and related components must be visually checked. This includes checking all fluid levels, finding

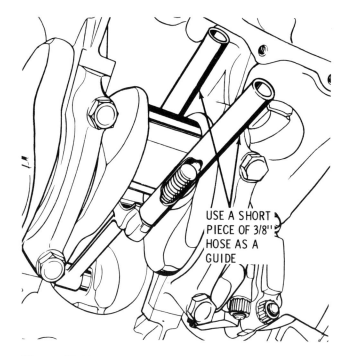

Figure 12-5 Protecting the crankshaft journal (*Courtesy of Oldsmobile Division*)

any leaks, and noting physical damage. A record on the service history of the vehicle may be helpful toward finding a solution.

The compression test is a quick, accurate way of determining the engine's physical condition. The compression test on the diesel is very similar to the compression test performed on the gasoline engine. Extra care is needed, however, because the cylinder pressures are much higher. Therefore, only a compression gauge designed for diesel engines should be used.

Engine removal and disassembly should be performed only when necessary and with the proper tools, equipment, and information. The components of the diesel engine are heavier than the comparable components found on the gasoline engine. Only equipment designed for the extra load should be used, and caution is paramount when physically moving the parts. It is essential that the technician mark and note the position of each component before it is removed.

CHAPTER 12 QUESTIONS

1. Why are visual checks important?
2. Why must a diesel compression gauge be used?
3. Why is a wet compression test not performed on a high-CR diesel?
4. What precaution must be followed when washing the engine?
5. List four cautions to be observed when disassembling the engine.

chapter 13
The Engine Block, Piston, Piston Assembly, Crankshaft, and Main Bearings

Objectives

In this chapter you will learn:
- Four differences between the diesel and gasoline engine block
- Why a piston is matched and fitted to a particular cylinder bore
- The advantages of having a cylinder liner
- Two differences between a typical diesel piston and gasoline engine piston
- How the full-floating piston pin works
- Why the throw weights and journals of a diesel crankshaft are larger than those on a gasoline engine crankshaft
- Why some manufacturers selectively fit the main bearings
- How to service the engine block, piston, piston assembly, and crankshaft

THE ENGINE BLOCK

The engine block (sometimes called cylinder case) is the main structural member of the engine, figure 13-1. All the other components are either housed internally or are connected to it. The engine block not only accommodates these components, but is also designed to withstand the high pressures the diesel combustion process exerts on it. To withstand these loads, the engine block must be thicker, stronger, and of better quality material than the gasoline engine block. The cast iron may have carbon, silicon, and chromium to improve the elasticity and thermal expansion qualities. Extra webbing is used in the crankcase area to increase engine block rigidity.

To maintain good sealing qualities between the piston and block, and to extend the life of the engine, the block may have pistons *match fitted* to cylinder bores or use cylinder liners.

When the block comes from the foundry, not all cylinder bores are likely to be identical. To keep the proper clearance between the cylinder bore and piston, the pistons are made in different sizes. Then the piston is matched and fitted to a cylinder bore that will yield the correct clearance. There may be a code stamped on the piston and on the block deck next to the cylinder bore, figure 13-2. This is why it is so important when removing pistons to know which cylinder bore they came from. The piston and cylinder bore size can vary from cylinder to cylinder.

A cylinder liner is a steel sleeve pressed into the cylinder bore, figure 13-3. The advantage of this method is that if the liner is badly scored or worn, the old liner can be pressed out and a new one installed. The liner can then be honed or bored to achieve the proper cylinder-bore-to-piston clearance.

INSPECTION AND SERVICE OF THE ENGINE BLOCK

Inspecting the engine block requires the following:

- appropriate cleaning tools

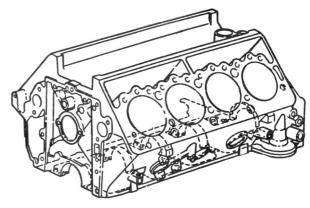

Figure 13-1 Engine block (cylinder case) (*Courtesy of General Motors Product Service Training*)

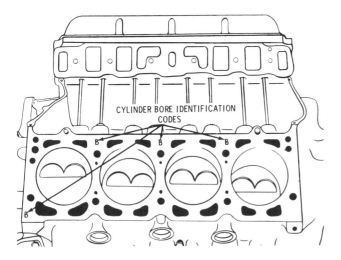

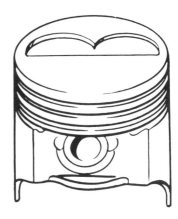

Figure 13-2 Cylinder bore and piston identification core indicate size of bore and piston. (*Courtesy of Oldsmobile Division*)

- straightedge and feeler gauges
- inside micrometer or cylinder bore gauge
- manufacturer's service manual

Inspect the engine block by following these procedures:

1. Thoroughly clean the engine block.
2. Inspect the engine block for cracks and defects in the block deck, cylinder walls, all passageways, camshaft and valve lifter bores, and main bearing webs.
3. Check the block deck surface for warpage with a straightedge. Compare to manufacturer's specification.

4. Measure the cylinder walls for wear, taper, and out-of-round, figure 13-4.
5. If measurements exceed specifications, the cylinder must be reconditioned or the liner replaced.

Service the engine block by following these procedures.

The boring and honing operation of a diesel engine block is virtually identical to the operations performed on the gasoline engine block. There are a few precautions to adhere to. If you must bore the block:

- An oversize head gasket may be necessary.
- The block deck should be free of debris. A file will usually remove any remaining dirt or burrs. Do this to ensure that the cylinder is at a right angle to the crankshaft when it is bored.
- The piston to be fitted should be measured with a micrometer, measuring at the center of the piston skirt and at right angles to the piston pin, figure 13-5. The cylinder should be bored to the same

Figure 13-3 Engine block with replaceable cylinder sleeve (*Courtesy of Ford Motor Company*)

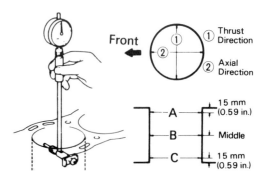

Figure 13-4 Measuring cylinder bore for taper and out-of-round

diameter as the piston and then honed to achieve the specified cylinder-bore-to-piston clearance.

- Always follow the manufacturer's instructions on boring and cylinder hone operation.

CYLINDER LINER SERVICE

On a block equipped with cylinder liners, minor scoring can usually be removed by honing. When the wear or damage is excessive, the liner must be replaced. This task requires the following:

- hydraulic press with appropriate adapters and the manufacturer's special tools
- dial indicator
- manufacturer's service manual

Remove and install the liner by following these procedures:

1. Secure the block to the press.
2. Install the adapters.

CAUTION: At this point, be sure that the block, press, and adapters are secure and will not interfere with anything. You will be applying approximately 7000 pounds of force to the liner. Damage to the block and personal injury may result if this procedure is not done properly.

---

3. Press out the liner, figure 13-6.
4. Inspect the liner bore for defects such as scoring.
5. Lubricate the outside surface of the new liner.
6. Press in the new liner and follow the previous caution.

Measure the protrusion of the cylinder liner above the block deck with the dial indicator. If protrusion is not

within specifications, remove the liner and install the correct liner shims, figure 13-7.

After pressing in the liner, bore and hone the liner bore to fit a standard piston.

THE PISTON, AND PISTON ASSEMBLY

Diesel engine pistons used in automobiles and light trucks look similar to gasoline engine pistons. The diesel piston is heavier because additional metal is needed to withstand the higher temperatures and pressures. The shape of the piston may vary, depending on type of combustion chamber and engine size.

Generally, the pistons are made of a cast aluminum alloy and are of a three-ring design. Some manufacturers have incorporated special features such as the following:

- A top ring groove insert. This material is much harder than aluminum. Cast iron and Ni-resist have been used, but Ni-resist is the preferred material. This special nickel alloy is extremely hard and has a high bonding-to-aluminum characteristic. These properties are necessary to prevent premature wear and to remain bonded because aluminum expands

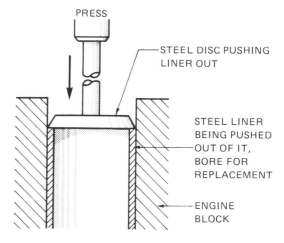

Figure 13-6 Pressing out the liner

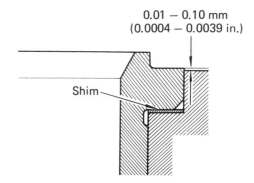

Figure 13-7 Use of shim to achieve specified sleeve protrusion

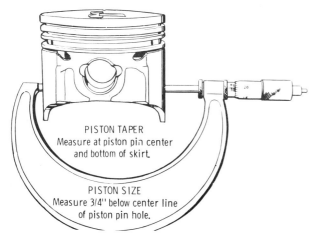

Figure 13-5 Measuring piston diameter (*Courtesy of Oldsmobile Division*)

and contracts quickly with temperature change, figure 13-8.

- The piston boss area is strengthened to handle the additional loads. This may be accomplished by adding more aluminum. A strut is included to control expansion, figure 13-9.
- The piston pin may not be offset as it typically is on gasoline engines. This reduces excessive loading on the piston.
- Grooves are used in the top land of the piston to break up carbon formations.
- The piston may be stamped, indicating that it is of a particular size and will fit properly only to the corresponding bore. A letter code is often used to indicate piston diameter.

Piston Rings

Like its gasoline counterpart, the diesel piston uses two compression rings to prevent combustion blow-by, and an oil control ring to scrape excess oil off the cylinder wall. The compression rings are often made of cast iron or high-strength iron and coated with either chrome or molybdenum. These coatings increase ring life and reduce cylinder scoring. Four types of compression rings are used—the compression or rectangular ring, the scraper ring, the inside bevel ring, and the keystone ring, see figure 13-10.

The compression ring and scraper ring are basically the same. But unlike the compression ring, the scraper ring has approximately a 1-degree taper, which allows it to seat quickly because of its initial line contact with the cylinder wall.

The inside bevel ring uses combustion gas pressure to force the bottom edge of the ring to contact the cylinder wall. Using combustion gas pressure helps the ring seal against blow-by.

The keystone ring is usually used in the top piston groove. Because of its shape, the force of compression

and combustion causes a certain amount of movement between the piston and the ring. This action frees the ring of carbon deposits and prevents sticking.

Though many different types of oil control rings are used, their specific purpose is to conform exactly to the shape of the cylinder wall to scrape off excess oil. A typical oil control ring consists of a thin top and bottom ring separated by an expander, figure 13-11.

Piston Pins

The piston pin is larger and thicker than the comparable gasoline engine piston pin. Three types of piston pins are the fixed, semifloating, and full-floating, see figure 13-12.

The full-floating type is the most common because of its ability to distribute the load and wear evenly. The full-floating piston pin concept is used to eliminate pin-to-boss scuffing and to promote uniform pin loading through pin rotation. The pin is held in by a spring clip at each end and is suspended by a thin film of oil. This permits the pin to rotate as the piston moves. The piston pin is lubricated by an oil passage through the connecting rod.

Connecting Rods

Connecting rods in diesel engines are usually forged rather than cast. On small diesel engines, the connecting rods are similar in appearance to those of gasoline engines;

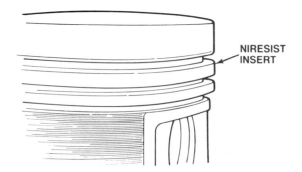

Figure 13-8 Hardened insert in top ring groove (*Courtesy of Ford Motor Company*)

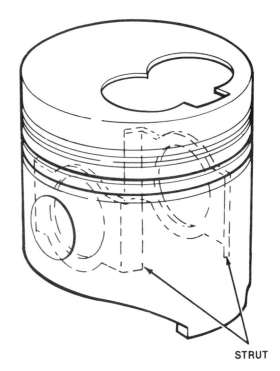

Figure 13-9 Struts are used to control expansion. (*Courtesy of Ford Motor Company*)

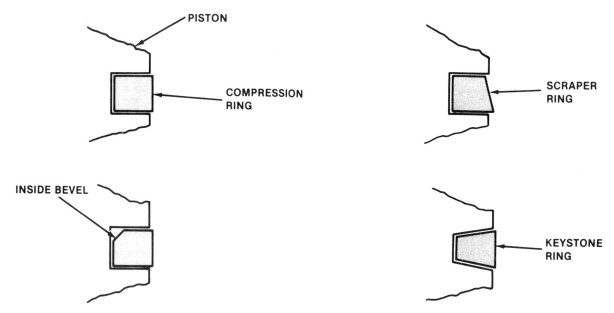

Figure 13-10 Different types of compression rings used on diesel engines *(Courtesy of Ford Motor Company)*

but as diesel engine size increases, the size of the connecting rods also increases to withstand the extreme loads placed on them. The connecting rod bearing is a precision insert type similar to the gasoline engine counterpart, though larger.

Piston, and Piston Assembly Service

Clean the varnish and carbon deposits with cleaning solvent.

CAUTION: Never use a wire brush for cleaning. Clean ring grooves with an appropriate groove cleaner and be sure all oil holes and slots are free of debris.

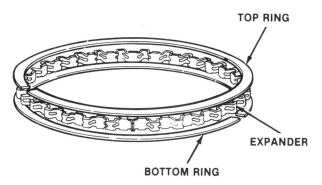

Figure 13-11 Oil control ring assembly *(Courtesy of Ford Motor Company)*

Inspect the piston for cracked ring lands, skirts, pin bosses, wavy or worn ring lands, damaged skirts, and excessive erosion across the top of the piston. Replace the piston if any of the above signs are detected, figure 13-13.

Measure piston diameter with a micrometer from the thrust faces (across center line of piston pin). If it is not within manufacturer's specifications, replace the piston.

Piston Pin. Check the pin fit by rocking the piston at right angles to the pin. If any freeplay is felt, the pin, piston, or both must be replaced.

CAUTION: After the pin is installed be sure the spring clips (retaining rings) are fully seated in their grooves by rotating the clips in their grooves.

Ring Gap and Installation. Install a compression ring in the cylinder bore to the depth specified by the manufacturer. Measure the space between the ring ends with a feeler gauge and compare to manufacturer's specifications, figure 13-14. Repeat this procedure for each cylinder.

Install piston rings with a piston ring expander.

CAUTION: Be sure to note which side of the ring the manufacturer wants facing upward. Roll each ring in its groove to be sure there is no binding or ring distortion. Measure the space between the ring

TYPES OF PISTON PINS

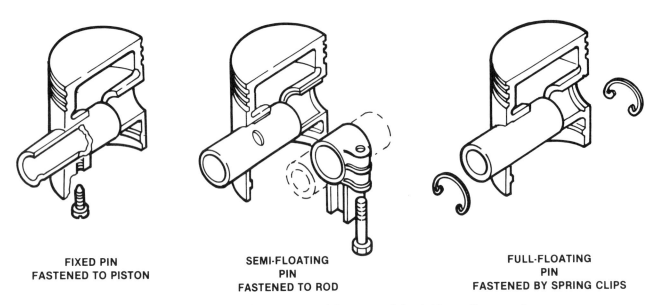

FIXED PIN
FASTENED TO PISTON

SEMI-FLOATING
PIN
FASTENED TO ROD

FULL-FLOATING
PIN
FASTENED BY SPRING CLIPS

Figure 13-12 Types of piston pins *(Courtesy of Ford Motor Company)*

Figure 13-13 Piston with part of the hardened insert missing

land and ring with a feeler gauge and compare to the manufacturer's specifications, figure 13-15.

Connecting Rod. Inspect the connecting rod and bearing surfaces for any signs of physical damage. Be sure the oil passageway is clear and free of debris. Replace the connecting rod if there are any defects.

CRANKSHAFT AND MAIN BEARINGS

The crankshaft of a diesel engine serves the same purpose as a crankshaft in a gasoline engine, converting linear motion to rotary motion. There are two major differences between them. The forging process for the

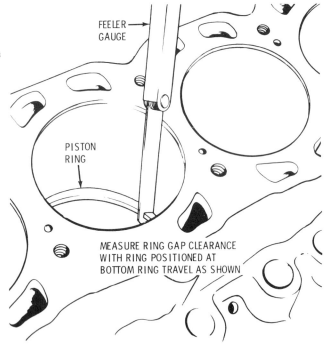

MEASURE RING GAP CLEARANCE
WITH RING POSITIONED AT
BOTTOM RING TRAVEL AS SHOWN

Figure 13-14 Measuring ring gap clearance *(Courtesy of Oldsmobile Division)*

diesel crankshaft is capable of producing a crankshaft that will withstand a greater load than the comparable cast crankshaft. Also, the diesel crankshaft is bigger. The journals are wider to withstand the greater stress of the heavier parts and pressure produced by combustion. To help maintain engine balance, the throw weights are heavier offsetting the greater weight of the piston and connecting

INSERT FEELER GAUGE AT
TOP OF RING GROOVE TO
MEASURE RING SIDE CLEARANCE

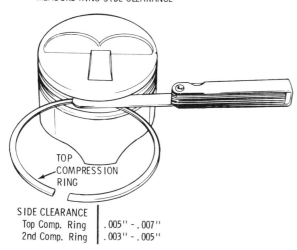

TOP
COMPRESSION
RING

SIDE CLEARANCE	
Top Comp. Ring	.005'' - .007''
2nd Comp. Ring	.003'' - .005''

Figure 13-15 Measuring ring groove side clearance (Courtesy of Oldsmobile Division)

rod, and the sudden acceleration load produced by the diesel combustion process.

The main bearing caps can be of a four-bolt or two-bolt design, figure 13-16. In either case, the main bearing caps are larger and wider than the comparable gasoline engine main bearings. The four-bolt design is preferred in larger automotive diesels because of its ability to withstand greater stress.

The main bearings are the precision insert type and may be selectively fitted by some manufacturers. Selective fitting means that the two insert halves are different sizes to maintain close tolerances. For example, one half of the bearing may be a standard insert and the other half a .001-in. undersize that will decrease the clearance by .0005 in. This is another reason it is important to know where each part came from.

Crankshaft and Main Bearing Service

To inspect a cranksahft you will need:

- a micrometer
- plastic gauging material
- a torque wrench
- manufacturer's service manual

First, visually inspect the crankshaft and bearings for any visible defects such as excessive wear or scoring.

Second, measure main journals, taking two readings at right angles from each journal. If wear is excessive, the crankshaft must be replaced or reground.

Third, measure the main oil clearance using plastic gauging material.

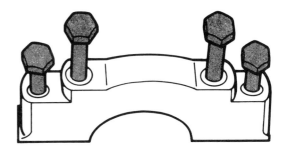

Figure 13-16 Four-bolt main bearing cap (Courtesy of Ford Motor Company)

1. Be sure the journal, cap, and bearing are clean.
2. Lay a strip of plastic gauging material across the full width of the journal.
3. Tighten the bearing cap bolts with a torque wrench according to the manufacturer's procedure and specifications.
4. Measure the plastigage at its widest. If the clearance is not within specifications, replace the bearings.

Be sure to follow any other procedure the manufacturer recommends for checking the crankshaft. If the crankshaft must be machined, be sure you know if there are replacement bearings available and how much the crankshaft can be safely machined.

SUMMARY

The diesel engine block, piston, piston assembly, crankshaft, and main bearings are made stronger to accommodate the increased demands placed on them by the diesel combustion process. Because of the need for strength in diesel components, construction, tolerances, and service may be different from a gasoline engine.

The diesel engine block has thicker cylinder walls and webbing. The cylinder bore and piston may be selectively fitted for each cylinder to maintain proper clearances. Cylinder liners may be used and have the advantage of being replaced when worn or damaged.

The piston and piston assembly are thicker and heavier than the comparable gasoline engine parts. The piston may have an insert in the top ring groove to resist wear and is reinforced in the piston pin and skirt portions. The piston pin is usually of the full-floating type to produce even wear and loading. The connecting rod is usually forged to handle extra stress.

The crankshaft and main bearings are larger and wider to disperse the increased load over a wider area. The crankshaft may be forged for extra strength. The main bearings are of the precision insert type and may also be selectively fitted to produce the proper clearances.

CHAPTER 13 QUESTIONS

1. List four differences between the diesel and gasoline engine blocks.
2. Why is a piston match fitted to a particular cylinder bore?
3. What are the advantages of using a cylinder liner?
4. List two differences between diesel and gasoline pistons.
5. Explain the full-floating piston pin concept.
6. Why are the throw weights and journals larger on a diesel crankshaft?
7. What is the purpose of using selectively fitted insert bearings?

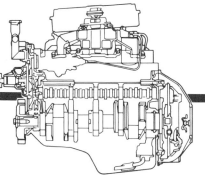

chapter 14
The Timing Drives, Camshaft, Valve Train, and Cylinder Head

Objectives

In this chapter you will learn:

● **The two different timing drives**
● **Four differences between a diesel and gasoline valve train**
● **Three differences between a diesel and gasoline cylinder head**
● **How to service the timing drive, camshaft, valve train, and cylinder head**

TIMING DRIVES

The timing drive is very important to the diesel engine. Not only must it drive the camshaft, but must also drive the injection pump at the proper time and speed. The timing drive must be very durable to handle the extra load imposed by the injection pump, the extra weight of the engine components, and the stress of the diesel combustion process. There are two basic types of timing drives—direct and indirect.

With *direct drive*, the camshaft is driven by gears, figure 14-1. Generally, a gear mounted on the crankshaft drives a gear mounted on the camshaft. On some engines, intermediate gears are used between the crankshaft, camshaft, and injection pump gears. This method is used when the camshaft is mounted in the block and the injection

pump is mounted in close proximity. The use of gears gives the timing drive high strength and long life. However, direct drive is costly and is difficult to use when the camshaft is mounted in the cylinder head, or if the injection pump must be mounted away from the engine block because of space limitations.

Indirect drive drives the camshaft by a chain or synthetic-toothed spur belt, figure 14-2. A roller-type chain (double row or single row) is used when the camshaft is either in the block or the cylinder head. The chain must be enclosed and constantly lubricated with engine oil.

The most popular method for driving an overhead camshaft in small diesel engines is the synthetic-toothed spur timing belt. The belt performs the same function as the chain but has the advantages of being quiet, light in weight, and relatively cheap. However, it is not considered

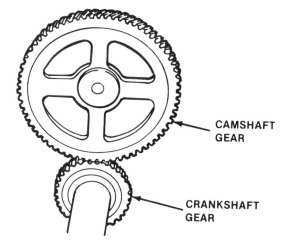

Figure 14-1 Direct-drive timing drive (*Courtesy of Ford Motor Company*)

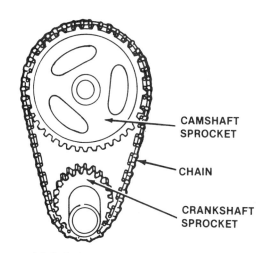

Figure 14-2 Indirect-drive timing drive (*Courtesy of Ford Motor Company*)

as durable as the other methods and may require periodic checking and service.

Servicing the Timing Drives

The service of each drive is different. It would be impractical to cover each in detail. Generally, any sign of extreme wear or physical defect is cause for replacement.

On direct drive, examine the gear teeth and contact pattern for defects. Gear teeth wear may be determined by the amount of backlash between the gears. This is done with a dial indicator mounted on the block, the stem against the camshaft gear. The amount of movement present is indicated on the dial indicator. This reading is compared to specifications, figure 14-3.

When a chain is used, the amount of wear is determined by the amount of freeplay present in the chain. This task is usually performed with a torque wrench and a ruler. With the torque wrench mounted on the camshaft, the torque wrench is turned in a specific direction to a specified value. The amount the chain moves is compared to the specifications and indicates whether replacement is necessary.

When a belt is used, any sign of deterioration is cause for replacement, figure 14-4. Belt tension is also critical and should be performed with the proper tools and service procedures. Regardless of the type of drive used, always determine the cause of failure before simply replacing the parts.

CAUTION: Always be certain that the timing marks are lined up properly. If it is not timed properly, the engine will be seriously damaged because there is very little clearance between the valves and piston under normal circumstances.

THE CAMSHAFT AND VALVE TRAIN

The camshaft performs the same function in the diesel engine as in the gasoline engine, namely, opening and closing the intake and exhaust valves at the proper time. The primary difference is that the camshaft in the diesel engine is made of harder materials to resist the increased friction. Harder cast iron or forged steel may be used to reduce wear. Friction is increased because of stronger valve springs, heavier valve train components, and contaminants in the engine oil (see Chapter 15).

When the camshaft is in the block, the valve train's major components are the lifter (tappet), pushrod, rocker arm, valve and valve assembly, figure 14-5.

There are two basic types of lifters used in small diesels, the mechanical and the hydraulic.

The mechanical (solid) lifter is the simplest type. To reduce friction and wear, the surface riding against the camshaft may be flattened (mushroomed). The mushroomed area disperses the load over a wider area reducing wear, figure 14-6. Mechanical lifters must also be adjusted

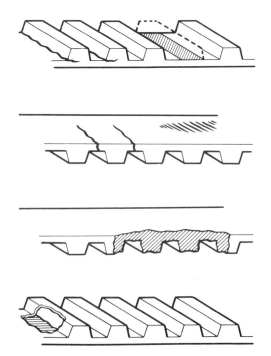

Figure 14-3 Measuring gear teeth wear (Courtesy of Ford Motor Company)

Figure 14-4 Replace belt if showing signs of deterioration.

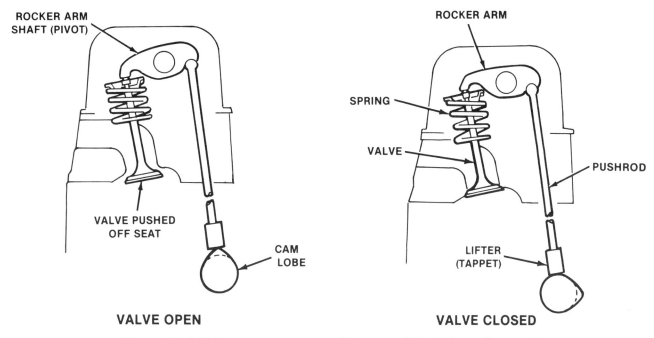

Figure 14-5 Valve train components (*Courtesy of Ford Motor Company*)

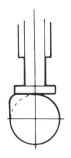

Figure 14-6 Mushroom lifter (*Courtesy of Ford Motor Company*)

periodically since there is no self-adjusting mechanism to compensate for wear.

Hydraulic lifters in the diesel operate the same way as in a gasoline engine, maintaining zero valve lash through the use of engine oil and a check valve, figure 14-7. The hydraulic lifter is made of a harder steel and may also be oversized. Because of production tolerances, the lifter bore may be larger than normal, so an oversized lifter is fitted into that bore. The bore and lifter are usually marked, indicating that they are oversized, figure 14-8. This is why it is important to return the lifter to the proper bore.

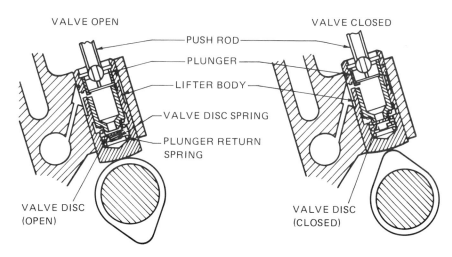

Figure 14-7 Hydraulic lifter operation (*Courtesy of Oldsmobile Division*)

To extend oil change intervals and decrease wear, several manufacturers are now using roller lifters or roller rocker arms. The roller revolves on a needle bearing, thus reducing wear between the lifter and camshaft, figure 14-9. However, the roller lifter must remain parallel with the camshaft to prevent severe damage. Therefore, guides are used to prevent the roller lifter from turning in its bore, figure 14-10. The roller rocker arm is used with some overhead cam engines to reduce friction between the rocker and valve, figure 14-11.

Pushrods are strengthened and the tips hardened to handle the increased stress. When roller lifters are used, only the upper end of the pushrod is hardened since the lifter no longer rotates. It is critical that the pushrod be marked to identify which end is up (to prevent premature wear of the pushrod).

Rocker arms are supported and operated by conventional methods and are usually larger to handle the extra load imposed by the larger valve springs. The rocker arm may also provide a means for adjusting valve lash.

When the camshaft is mounted in the cylinder head, lifters, pushrods, and in some cases, rocker arms are not used. Eliminating these components allows the engine to be more efficient and operate at higher RPM because there is less mass to move. With the overhead camshaft, adjustment is accomplished by changing shims or turning a screw mounted in the rocker arm, figure 14-12.

Intake and exhaust valves are subjected to greater abuse on a diesel. They must provide an extremely tight seal against the high combustion pressures, resist the high temperatures, carbon buildup, and the impact caused by higher valve spring pressures. To accomplish these tasks, the valves are made of special steel alloys that are harder and have a higher melting temperature than conventional valves. The valves are also larger and have a wider margin to resist the high temperatures and increase strength.

Some exhaust valves use sodium to keep them cool. The valve stem is hollow and partially filled with sodium. When the sodium melts (approximately 400°F [202°C]), it moves back and forth within the stem, transferring heat from the lower region of the valve to the upper stem, where it is dissipated into the cylinder head, figure 14-13.

To ensure a proper fit between the valve guide bore and stem, some manufacturers install a valve with an oversized stem. The valve stems come in different oversizes, as needed. The valve stem seals also come in different sizes indicated by the color of the seal.

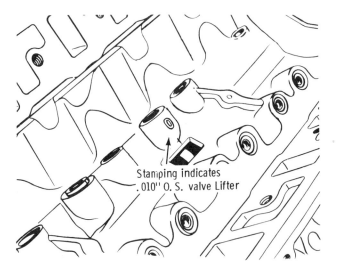

Stamping indicates .010" O. S. valve Lifter

Figure 14-8 Oversize lifter bore (*Courtesy of Oldsmobile Division*)

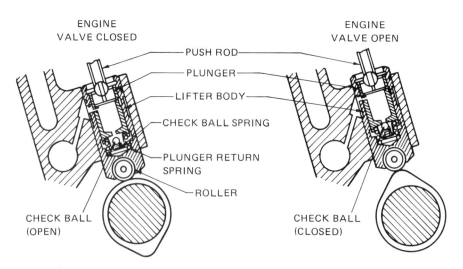

Figure 14-9 Roller lifter operation (*Courtesy of Oldsmobile Division*)

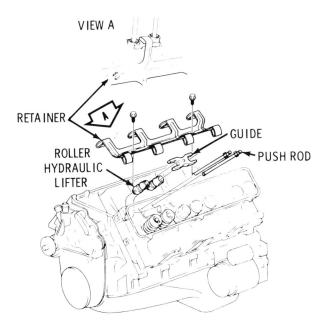

Figure 14-10 Retainers and guides prevent the roller lifter from turning. *(Courtesy of Oldsmobile Division)*

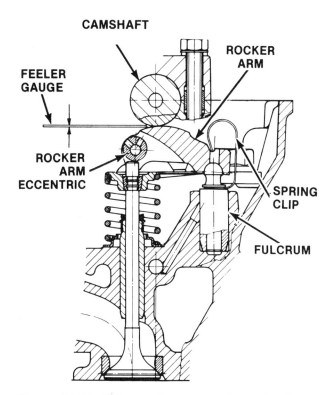

Figure 14-11 Roller rocker arm with overhead camshaft. *(Courtesy of Ford Motor Company)*

High-strength, high-pressure valve springs are needed to ensure that the valve forms a tight seal against the valve seat and that the valve does not float at high engine RPM. Remember that the valve train components on a diesel are heavier and require a higher-pressure spring.

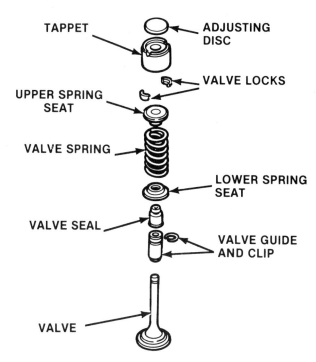

Figure 14-12 Changing adjusting disc thickness changes the clearance between camshaft and disc. *(Courtesy of Ford Motor Company)*

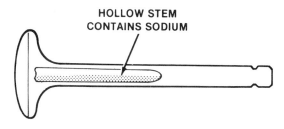

Figure 14-13 Exhaust valve, sodium filled for added cooling *(Courtesy of Ford Motor Company)*

To reduce valve seat wear, hot spots, and carbon buildup, valve rotators are used. This device is mounted on top or under the valve spring and rotates the valve slightly, each time the valve moves.

Servicing the Camshaft and Valve Train

The camshaft and bearings are serviced the same as in a gasoline engine. Common checks are camshaft end play, camshaft runout, and cam lobe height.

Solid and hydraulic lifters should be checked and measured for excessive wear and scoring. With hydraulic lifters, follow the manufacturer's service procedure for disassembly and cleaning.

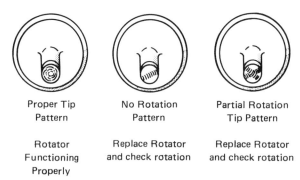

Figure 14-14 Valve wear patterns *(Courtesy of Oldsmobile Division)*

CAUTION: Be certain to keep each lifter separate from the others and to know which bore they go in.

When a roller is used, visually check the roller for signs of scoring or pitting, and missing or broken needle bearings. The roller should rotate freely without binding or excessive play. If any of these conditions is observed, the lifter must be replaced and the corresponding camshaft lobe checked for wear. Check the pushrods for wear on each end; be sure the oil passageway is clear and the pushrod is not bent. Be certain you know which way the pushrod is to be installed.

Check the rocker arms, pivot studs, or shaft for signs of wear or scoring, and replace if necessary.

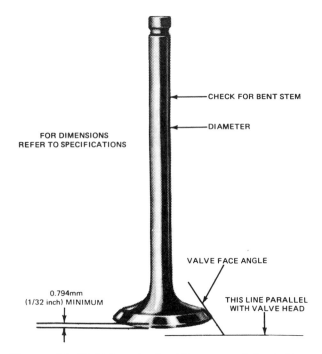

Figure 14-15 Valve checks *(Courtesy of Ford Motor Company)*

Check valve spring squareness and pressure, according to manufacturer's procedures.

Check if the valve rotators have been working by inspecting the tip of the valve stem for the proper pattern, figure 14-14. If the rotator is not functioning, replace it.

Inspect the valve margin for the proper dimension. If it is too thin, replace the valve, figure 14-15.

If the valve has an adequate margin and is not physically damaged, it can be refaced. Follow the manufacturer's instructions on refacing the valve and grinding the stem tip. Also, check the service manual for the specified valve angle.

Valve stem height must be measured after the valve and valve seats are ground, and whenever a new valve is installed. This task requires the following:

- manufacturer's gauge tool
- carburetor gauge or feeler gauge
- manufacturer's service manual

Measure valve stem height by the following procedures:

With the valve installed, set the special tool in the specified area on a clean, smooth surface. Measure the gap between the special tool and valve stem, figure 14-16. If the clearance is less than specified, grind the valve stem to the proper specifications. This clearance is necessary to ensure that the rocker arm does not interfere with the rotator. Next, measure the gap between the rotator and valve stem tip, figure 14-17. If it is less than specified, replace the valve.

THE CYLINDER HEAD

The diesel cylinder head is subjected to high pressure, high temperature, and thermal shock (sudden temperature changes), particularly when the engine is first started. To withstand this abuse, the diesel cylinder head is made of premium materials and is larger than the comparable gasoline cylinder head. The head may be made of cast iron or aluminum, though cast iron is currently more popular. Iron is more durable and retains heat better, which is important when operating under light load conditions. The disadvantage is that it is heavy. Aluminum is light in weight but transfers heat easily, making heat retention harder under light load operation. To help hold the cylinder head in place, there are generally more cylinder head bolts surrounding each cylinder. They are larger and usually torqued to a higher value. A new method is being used on some diesel engines to supplement the use of a torque wrench when tightening the head bolts. With this new method these bolts are torqued to a specified value, then rotated a specific amount, putting the proper amount of pressure on the cylinder head. For example, the bolts are torqued then rotated one quarter

turn, twice, figure 14-18. This stretches the bolt and eliminates the need for later retightening of the head bolts. It is critical that the bolt be measured for the proper length before installation, figure 14-19.

High-grade steel or stainless steel inserts make up the antechamber area, particularly the lower half. These steels are used to resist the eroding effects of the hot combustion flame exiting the antechamber area. Because they are made in the form of inserts they are replaceable when worn, figure 14-20.

The valve guide bores may be oversized because of slight imperfections when they are made. A number is usually stamped on the cylinder head near the valve guide bore indicating the amount of oversize, figure 14-21. Naturally, an oversized valve stem would be required to fit properly.

The valve seat area is another place that takes extreme abuse. Because the valves are hardened to resist wear and provide good sealing, the valve seat area is also hardened. Valve seats are either induction hardened or may use special steel inserts, figure 14-22. The inserts have the advantage of being replaceable when excessively worn. Hardening the valve seat area also maintains the

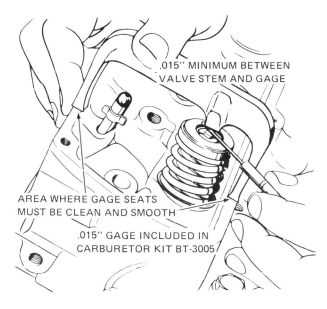

.015″ MINIMUM BETWEEN VALVE STEM AND GAGE

AREA WHERE GAGE SEATS MUST BE CLEAN AND SMOOTH

.015″ GAGE INCLUDED IN CARBURETOR KIT BT-3005

Figure 14-16 Valve stem height check (*Courtesy of Oldsmobile Division*)

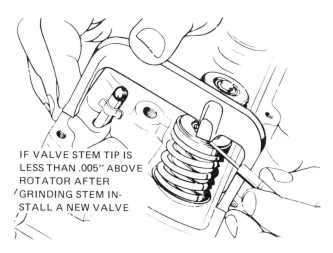

IF VALVE STEM TIP IS LESS THAN .005″ ABOVE ROTATOR AFTER GRINDING STEM INSTALL A NEW VALVE

Figure 14-17 Measuring distance from rotator to valve stem tip (*Courtesy of Oldsmobile Division*)

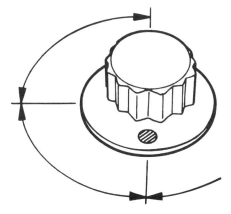

Figure 14-18 Mark on head bolt helps to identify how many degrees the bolt has turned. (*Courtesy of Ford Motor Company*)

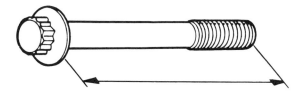

Figure 14-19 Head bolt length is critical. Bolts not meeting the specified length should be discarded. (*Courtesy of Ford Motor Company*)

FULL LOAD
ENGINE SPEED: 4350 1/min
TEMPERATURES IN °C

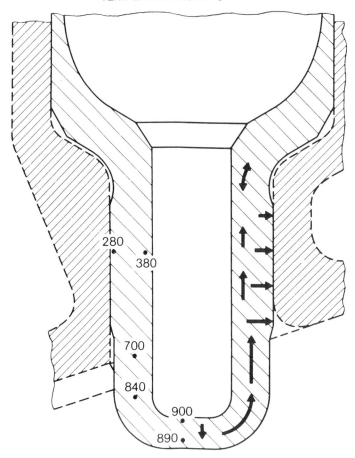

Figure 14-20 Maximum temperatures of lower prechamber and heat flow pattern *(Courtesy of Mercedes-Benz of North America Inc.)*

Figure 14-21 Marking indicates amount of valve guide oversize. *(Courtesy of Oldsmobile Division)*

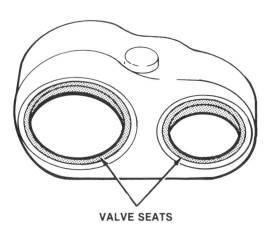

Figure 14-22 Valve seat inserts *(Courtesy of Ford Motor Company)*

correct valve seat width. If the seat becomes too great, too much heat would be conducted away from the valve, allowing carbon buildup. If the seat became too narrow, not enough heat would be conducted away and the valve would burn.

Cooling the diesel cylinder head is important, particularly in critical areas. Coolant passages are cast around the antechambers and valve guides, since these are the hottest areas. Casting a diesel cylinder head is harder for the manufacturer because of the close tolerances and intricate coolant passageways.

The diesel cylinder head gasket has a hard life. It must effectively seal against high pressures and sudden thermal shocks. Also, the ratio of cylinder head bolt torque to combustion pressure is low when compared to a gasoline cylinder head. For example, a gasoline 350-cu.-in (5.7-L) engine has a cylinder head bolt torque specification of 65 ft/lb (89 N·m), and under normal conditions, cylinder compression is approximately 120–140 psi (827–965 kPa). The comparable diesel engine has cylinder head bolt torque of 130 ft/lb (176 N·m), and under normal conditions cylinder compression is approximately 370–400 psi (2251–2758 kPa). In this example, cylinder compression on the diesel is three times larger than cylinder head bolt torque specification. The gasoline engine is only twice as large. To cope with these adverse factors, the head gasket is made with special materials and sealers. The sealer is printed in the proper amount on the gasket to ensure good sealing.

Servicing the Cylinder Head

Diesel cylinder head service is similar to that for gasoline cylinder heads, with several extra operations and cautions. Thoroughly clean the cylinder head.

CAUTION: Do not damage the cylinder head gasket surface and be very cautious with aluminum cylinder heads.

Variations of the cylinder head gasket surface can affect the sealing quality and even the compression ratio. Because the compression ratio is so high, minor variations can change CR between cylinders.

After visually inspecting for physical damage, check the cylinder head gasket surface for warpage with straight-edge and feeler gauge. Compare the readings to manufacturer's specifications. If the head is warped, it must be replaced. Resurfacing the head would increase the CR above engine tolerances.

Check the valve-stem-to-guide clearance. If the valve guide bore is worn excessively, it must be replaced, knurled, or reamed oversized to the proper specifications.

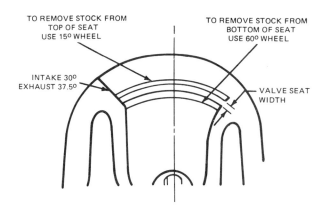

Figure 14-23 Valve seat width must be correct for good valve seating. *(Courtesy of Ford Motor Company)*

Grind the valve seats to the proper angle and width according to the manufacturer's instructions, figure 14-23.

CAUTION: First, be certain to know the proper valve seat angles. Second, a valve seat grinding stone must be dressed frequently because of the hardened valve seats.

As stated earlier, some manufacturers use valve seat inserts. This task requires the following:

- appropriate hand tools
- manufacturer's special tools
- manufacturer's service manual

Install valve seat inserts by following these procedures: Install the special seat-removing tool according to the manufacturer's instructions and remove the seat insert, figure 14-24. If an oversized seat insert is needed, the insert bore must be counterbored.

A new valve seat insert can be installed using the appropriate special tool and a hammer. Tap the insert until it is seated, figure 14-25.

Inspect the antechamber and throat area for excessive wear. Replacing most antechambers requires the following:

- narrow drift
- ball peen hammer
- 1-in micrometer

Replace the antechamber by following these procedures: Insert the drift either through the glow plug or nozzle hole and tap out the antechamber, figure 14-26.

CAUTION: Be careful not to damage the nozzle seat or threads. Remove any remaining carbon in the antechamber.

Measure the height of the original antechamber and grind off any excess found on the replacement chamber, according to manufacturer's instructions.

Note that the antechamber has a notch and can be installed in only one direction, figure 14-27. Tap in the new antechamber to the desired dimension. If the ante-chamber is protruding too far, the cylinder gasket will not seal.

With certain manufacturers, the thickness of the head gasket must be determined. This is done by measuring the amount the piston protrudes above the engine block, figure 14-28. The amount of piston protrusion determines the proper head gasket. This is critical in maintaining correct compression ratio and valve clearances.

SUMMARY

Timing drives for diesels are either direct or indirect. Not only do they drive the camshaft, but also the fuel injection pump. The timing drive in a diesel must be stronger to handle the extra stress imposed by the diesel process and injection pump.

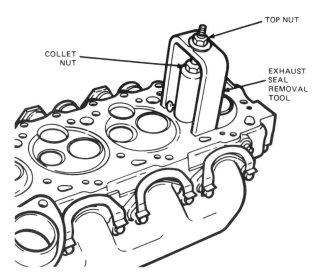

Figure 14-24 Removing valve seat insert (*Courtesy of Ford Motor Company*)

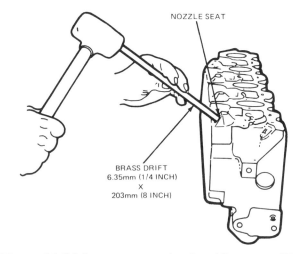

Figure 14-26 Removing antechamber (*Courtesy of Ford Motor Company*)

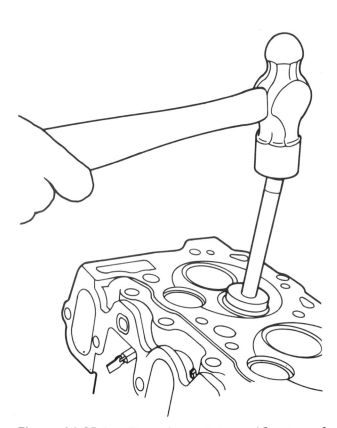

Figure 14-25 Installing valve seat insert (*Courtesy of Ford Motor Company*)

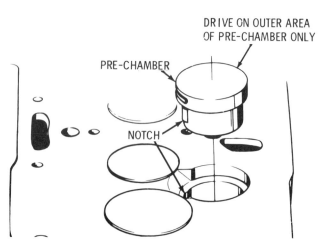

Figure 14-27 Installing antechamber (*Courtesy of Olds-mobile Division*)

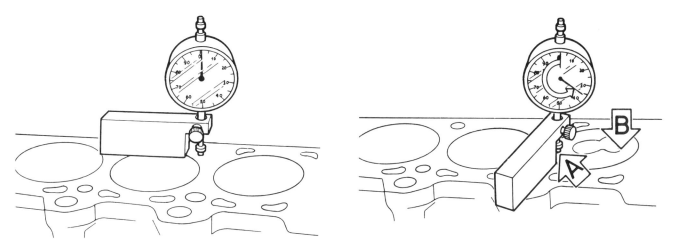

Figure 14-28 Measuring piston protrusion with dial indicator *(Courtesy of Ford Motor Company)*

The camshaft and valve train are made of harder, stronger materials to resist the increased friction and loads. Other methods include using mushroom or roller lifters to decrease friction.

The valves are larger and heavier to withstand the heat and pressure exerted on them. Valve sealing is critical in a diesel in maintaining high compression pressures. Valve seat width is critical in keeping the valve cool enough to prevent burning, but also in preventing carbon buildup. Some exhaust valves may use sodium to increase valve cooling. To maintain proper valve-stem-to-valve-guide-bore clearance, some manufacturers use oversized valve stems. If the valve guide bore is oversized, the number indicating its size is stamped near it.

Diesel cylinder heads are built to withstand the diesel combustion process and thermal shock. There are usually more and larger bolts holding the cylinder head in place. Coolant passages are cast around critical areas such as the antechamber and valve guide area.

Diesel cylinder head service is similar to gasoline cylinder head service. However, greater caution is needed when servicing the diesel cylinder head because of the very close tolerances and high pressures involved.

CHAPTER 14 QUESTIONS

1. Describe the two different timing drives used on diesel engines.
2. List and explain four differences between a diesel and gasoline engine valve train.
3. Why must the lifters be installed in their original bores?
4. Why are roller lifters used?
5. List three differences between a diesel and gasoline cylinder head.
6. Why, with certain tightening methods, must the cylinder head bolt length be measured?
7. Why can a diesel cylinder head not be machined?

chapter 15
The Lubrication and Cooling Systems

Objectives

In this chapter you will learn:
- The operation of the lubrication and cooling systems
- Why oil coolers are used
- The different oil classifications
- Why piston cooling jets are used
- Why the diesel has a larger lubrication and cooling system
- How soot and sulfur get into the engine oil and why they are detrimental to the engine
- Why some cooling systems need a high ratio of ethylene glycol to water
- How to service the lubrication and cooling system

THE LUBRICATION SYSTEM

The purpose of the lubrication system is to distribute oil to key areas throughout the engine. Major lubrication components used on diesel engines are the crankcase sump pan, oil pump with pressure regulator, oil pump drive mechanism, oil cooler, oil cooler bypass valve, oil filter, oil filter bypass valve, oil pressure sending unit, and the piston-cooling jets.

With the exception of the oil cooler and piston-cooling jets, the lubrication system components of the diesel engine are similar in design and operation to those of the gasoline engine. Typically, oil is pulled through a screened inlet near the bottom of the crankcase sump pan. The pressurized oil passes through the oil filter and then to the oil cooler. From the oil cooler the oil enters the main oil gallery. This gallery usually feeds the crankshaft and connecting rod bearings. From there the oil is distributed to smaller galleries that feed such components as the lifters, camshaft bearings, timing drive, valve train assembly, vacuum pump drive, and piston-cooling jets, figure 15-1.

The oil capacity of the diesel lubrication system is generally 1 to 2 guarts higher than that of the comparable gasoline engine. The extra volume is needed to support the larger bearing surfaces and assure adequate oil flow through the oil cooler. To supply enough oil at the proper pressure to these key areas, a gear or trochoid-type oil pump is used that has larger pumping capacity than the comparable gasoline engine oil pump, figures 15-2, 15-3.

Oil coolers are devices in which the hot oil gives up its heat to the circulating engine coolant. This helps maintain the oil below the temperature at which it would oxidize and deteriorate. Oil coolers are mounted in one of three places—in the radiator, in the engine block, or bolted onto the engine block.

Regardless of the location, the operation of the oil cooler is the same. Oil and engine coolant flow in separate passages, allowing the engine coolant to take heat away from the hot oil. Should the oil cooler become plugged, a bypass valve would open, carrying oil to the engine, figure 15-4.

Also found on some diesel engines are piston-cooling jets. These are nozzles that direct an oil spray to the underside of the piston, figure 15-5. The oil conducts heat away from the piston. This not only preserves the life of the piston, but allows the use of a lighter piston. Without the piston-cooling jets, the pistons would need more metal to absorb and conduct heat away.

A pressure check valve may be incorporated just before the piston-cooling jet, figure 15-6. This valve opens only when the oil pressure is high enough, and when the extra cooling is needed at high RPM. This ensures that other engine components receive the proper amount of oil at low engine speeds.

Engine Oil

As in the gasoline engine, the oil must provide good fuel economy and protect the diesel engine from wear, rust, and harmful chemicals. In addition, the oil must protect the diesel engine from premature wear caused by soot (carbon) and acids. The diesel engine requires an oil that has the proper additives to inhibit carbon wear and

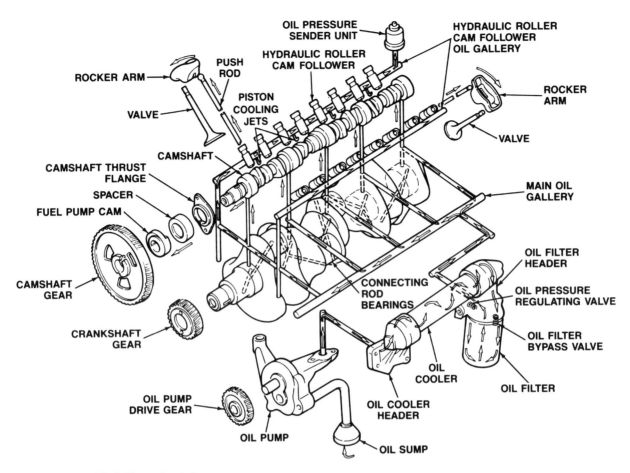

Figure 15-1 Typical oil flow through the lubrication system *(Courtesy of Ford Motor Company)*

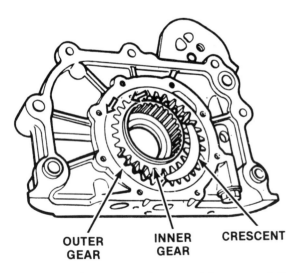

OUTER GEAR   INNER GEAR   CRESCENT

Figure 15-2 Gear-type oil pump *(Courtesy of Ford Motor Company)*

acid formation. Oil companies have formulated oils that meet the needs of both gasoline and diesel engines in passenger cars and light trucks. The technician must know the oil classifications pertaining to both gasoline and diesel engines.

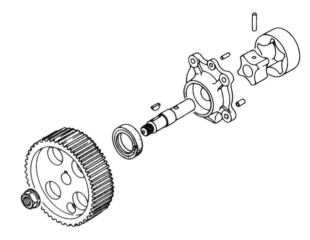

Figure 15-3 Trochoid oil pump *(Courtesy of General Motors Product Service Training)*

The *viscosity* (resistance to flow) of the oil is measured at certain temperatures by the SAE (Society of Automotive Engineers) and assigned a number. Chemical additives are added to change the viscosity of the oil with engine temperature, figure 15-7. A multi-viscosity oil can be thinner at lower temperatures to provide easier starting,

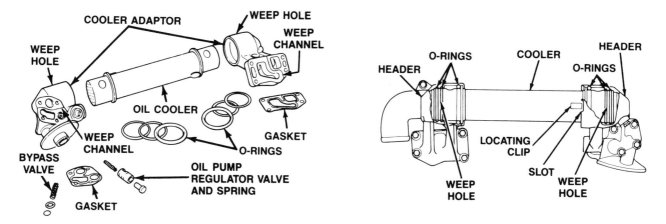

Figure 15-4 Engine oil cooler *(Courtesy of Ford Motor Company)*

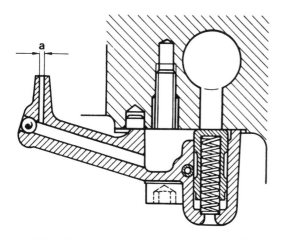

Figure 15-6 Piston-cooling jets *(Courtesy of Mercedes-Benz of North America Inc.)*

and changes viscosity as engine temperature increases, providing good lubrication. A typical viscosity rating would be 10W-40.

The API (American Petroleum Institute) service categories are divided into two major areas. The S service category is used for gasoline engines, specifically to denote service station oils or spark ignition engine oils. The second letter ranges from A to F. SA rating provides the least protection, SF the greatest protection, figure 15-8.

The second major API service category begins with a C, used to denote commercial oils or compression-ignition engine oils. Again, the second set of letters describes the oil's performance, ranging from CA to CD, with CA providing the least protection and CD providing the greatest protection, figure 15-9. These API oil ratings are used in determining the proper oil for large diesel engines.

With the advent of passenger-car and light-truck diesels, oil companies provided oil that meets both gasoline and diesel engine requirements. Therefore, when choosing the

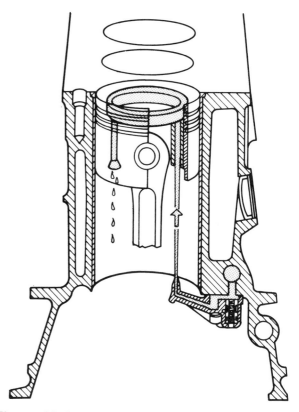

Figure 15-5 Engine oil sprayed into the piston to conduct heat away *(Courtesy of Mercedes-Benz of North America Inc.)*

proper oil for a passenger car diesel, it must be of THE PROPER VISCOSITY AND MEET BOTH API SERVICE CATEGORIES. For example, it is possible to buy a 10W-30 SE-CC and 10W-30 SE-CD oil. Both oils provide different levels of protection. Consult the manufacturer's service manual for the proper oil.

The diesel engine imposes special lubricating oil requirements. The diesel combustion process produces carbon (soot) which is absorbed into the oil. If the oil remains

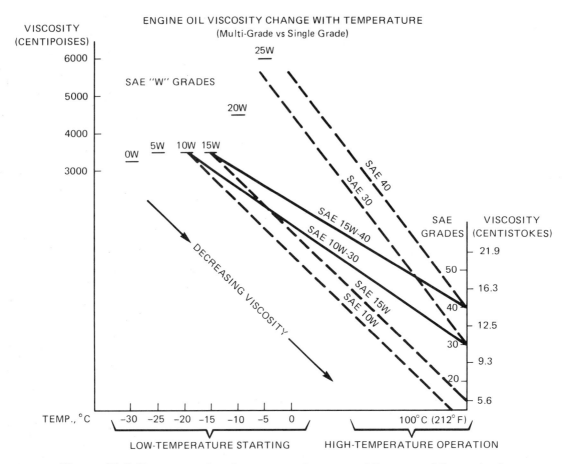

Figure 15-7 Temperature's influence on oil viscosity *(Courtesy of Exxon Inc.)*

SERVICE STATION OILS

API SERVICE CATEGORY	OIL DESCRIPTION
SA	STRAIGHT MINERAL OIL (NO ADDITIVES
SB	ANTI-OXIDANT, ANTI-SCUFF, BUT NON-DETERGENT*
SC	PROTECTION AGAINST HIGH- AND LOW-TEMERATURE DEPOSITS, WEAR, RUST, CORROSION; MEETS CAR MAKERS' WARRANTY REQUIREMENTS FOR 1964-1967 MODELS.
SE	IMPROVED PROTECTION OVER SD OILS; MEETS WARRANTY REQUIRE-MENTS FOR 1972-1980 MODELS.
SF	IMPROVED ANTI-WEAR AND ANTI-OXIDATION; MEETS WARRANTY RE-QUIREMENTS FOR 1980 AND LATER MODELS.

*NOT RECOMMENDED FOR USE IN ENGINES UNLESS SPECIFI-CALLY RECOMMENDED BY THE MANUFACTURER.

Figure 15-8 API service categories for gasoline engines *(Courtesy of Exxon Inc.)*

COMMERCIAL OILS

API SERVICE CATEGORY	OIL DESCRIPTION
CA	LIGHT-DUTY SERVICE; MEETS OBSOLETE* MILITARY SPECIFICATION MIL-L-2104A.
CB	MODERATE DUTY; MEETS OBSOLETE* MIL-L-2104A, SUPPLEMENT 1.
CC	MODERATE-TO-SEVERE DUTY; MEETS OBSOLETE* MIL-L- 2104B.
CD	SEVERE DUTY; HIGHEST PROTECTION AGAINST HIGH- AND LOW-TEMPERA-TURE DEPOSITS, WEAR, RUST, COR-ROSION; MEETS MIL-L-2104C.

*OBSOLETE FOR MILITARY USE, BUT REMAINS VALID DE-SCRIPTION OF API SERVICE CATEGORY.

Figure 15-9 API service categories for diesel engines *(Courtesy of Exxon Inc.)*

in the engine longer than the manufacturer specified, the carbon particle content causes *carbonaceous wear* (wear by carbon particles). Engine components under high load, like the camshaft and lifter surfaces, are particularly susceptible to carbonaceous wear. Switching to roller lifters allows the manufacturers to increase the mileage between oil changes.

Diesel fuel is higher in sulfur content than gasoline. The sulfur combines with water to form acids that eat away at the engine's internal parts. Additives are needed to reduce the acidity of the oil, but these wear out. They also wear out earlier if the sulfur content of the fuel is high.

Because of the sulfur in the fuel and the carbon produced by the diesel combustion process, the oil change intervals are more frequent on a diesel engine.

The viscosity of the oil is also critical. If the viscosity is too low, the oil does not provide adequate lubrication at high temperatures. If the viscosity of the oil is too high, the engine starts hard and the piston rings stick, causing excessive blow-by. Follow the manufacturer's recommendations on viscosity.

Servicing the Lubrication System

Thoroughly clean all parts of the lubricating system. Flush out the internal passages of the oil cooler and the oil cooler lines. Look for signs of contamination and metal particles and determine the reason for their presence.

Disassemble the oil pump according to the manufacturer's instructions. Inspect the internal parts for signs of wear. Usually, the oil-pump body and gear faces are checked with a straightedge and feeler gauge in the same manner as a gasoline engine oil pump, figure 15-10. If the pump does not meet specifications, or you are in doubt on its condition, replace it.

Carefully inspect the oil pump drive. When the oil pump is gear driven, check the tooth contact pattern and backlash. If the oil pump is driven by a shaft, check the shaft for scoring, rounded edges, and see if it is twisted or bent. If any of these conditions are, or seem to be present, replace the oil pump drive. It is much cheaper than replacing the engine.

Check the pressure regulator and bypass valves for signs of scoring and wear. The valves should also slide freely without sticking in their bores.

If possible, inspect the oil cooler for signs of corrosion and leakage. Check the gasket surfaces and line fittings. Check the oil cooler lines for corrosion and restrictions. Look in areas where the lines may be rubbing against other components. Flush the oil cooler to remove any metal particles or debris that may have collected in the passages.

On engines equipped with piston-cooling jets, the cooling jets may have been removed to allow for adequate clearance. Installing the piston-cooling jets requires the following:

- the manufacturer's installation and alignment tools
- hammer
- manufacturer's service manual

Install the piston-cooling jets by following these procedures:

Set the piston-cooling jet and installation tool according to the manufacturer's instructions. Lightly tap on the tool to start the jet into its bore, figure 15-11.

Tap on the installation tool until the jet is seated.

It is critical that the oil spray coming from the piston-cooling jet be correct to ensure piston cooling. Mount the alignment tool according to the manufacturer's instructions. The piston-cooling jet should line up with the appropriate

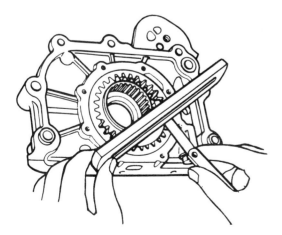

Figure 15-10 Oil pump side clearance (*Courtesy of Ford Motor Company*)

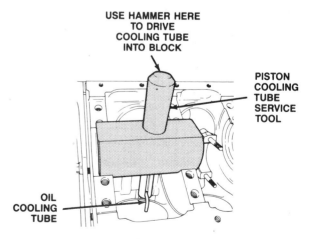

Figure 15-11 Installing piston-cooling tube (*Courtesy of Ford Motor Company*)

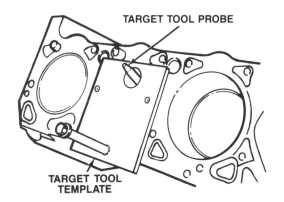

Figure 15-12 Checking piston-cooling tube alignment *(Courtesy of Ford Motor Company)*

target, figure 15-12. If it does not, the piston-cooling jet can be bent slightly until it is in alignment.

THE COOLING SYSTEM

Diesel engine cooling systems in passenger cars and light trucks are very similar to gasoline engine cooling systems in construction and operation, figure 15-13. The purpose of the cooling system is to allow the engine to warm up quickly and maintain the most efficient operating temperature without causing engine damage.

When the engine is cold, coolant is circulated through the coolant passages in the engine by the water pump. The thermostat is closed, preventing coolant from circulating through the radiator, thereby bringing the engine to operating temperature faster. A bypass hose is used to keep coolant circulating when the thermostat is closed. The thermostat opens when the coolant reaches the proper temperature and allows the coolant to circulate through the radiator. There the coolant releases its heat to the surrounding air. Fluid-coupling fans or electric fans are used to increase the air flow over the radiator. They increase engine efficiency and economy since they work only when needed.

The diesel engine increases the load on the cooling system since most of them use an oil cooler. A larger radiator, greater cooling system capacity, and a larger or faster circulating water pump is needed to ensure proper coolant flow throughout the engine.

The coolant is made of ethylene glycol and water, commonly 50% of each. However, with the use of front

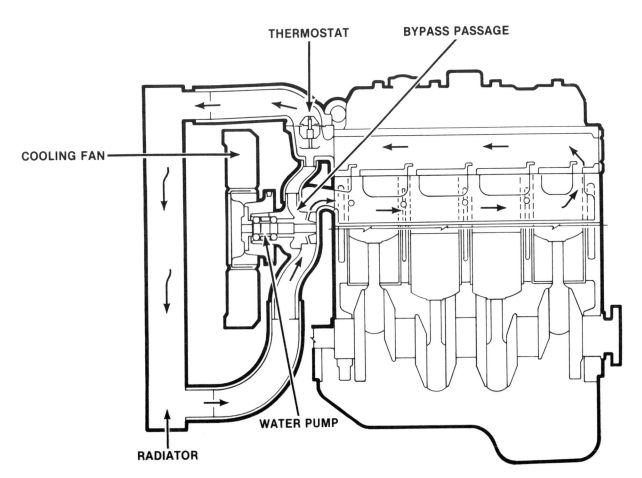

Figure 15-13 Coolant flow through the cooling system *(Courtesy of Ford Motor Company)*

wheel drive, aluminum heads, and smaller radiators, the coolant operating temperature is higher. To compensate, the ratio of ethylene glycol to water has been changed (generally to a 60% ethylene glycol, 40% water mixture) and a higher-pressure cap is installed, 16–18 psi (110–124 kPa). This raises the boiling point of the coolant, reducing the chance of boil-over under extremely hot conditions and high altitudes.

Cooling System Service

With the engine apart, inspect all cooling passages for scale and rust. It is important that scale and rust be removed to prevent hot spots.

Inspect all hoses for signs of rotting and deterioration. Replace as needed.

Inspect the radiator for physical damage and corrosion. Check the radiator mounts and support brackets.

Pressure check the radiator cap to see if it holds its specified rating.

Check the water pump bearing for excessive freeplay or roughness. Inspect the seal for leakage. Replace the bearing and seal, if defective.

Inspect the water pump impeller for looseness on the shaft and deteriorated impeller fins. If either condition is present, replace the pump.

Test the thermostat for proper operation by immersing the thermostat in water and gradually heating it. The thermostat should open at the specified temperature and close when cooled down. A common thermostat-opening temperature is 190–195°F (88–91°C).

If the fan has a fluid coupling, inspect damage and leaks. While rotating the coupling in your hand, there should be a definite resistance. If it is seized or rotates too freely, replace the coupling.

Check the fan for loose, bent, or broken blades.

CAUTION: Do not attempt to straighten a bent blade. That would increase the weakness possibly causing breakage and serious injury. If any of the above conditions are observed, replace the fan.

SUMMARY

The diesel lubrication system has a greater oil capacity, which is needed to supply the larger bearing surfaces and oil cooler. To ensure the oil gets there, the oil pump usually has a greater pumping capacity. The oil flow pattern is very similar to that of the comparable gasoline engine.

Oil coolers are needed to prevent the oil from deteriorating. This is accomplished by circulating engine coolant and oil in separate passages next to each other. Here heat from the oil is transferred to the engine coolant.

Piston-cooling jets spray oil to the underside of the piston. The oil conducts heat away from the piston, cooling the piston.

Oils used in passenger-car and light-truck diesel engines must be of the proper viscosity and meet both API service classifications. The oil must be formulated to inhibit carbonaceous wear and acid buildup due to the diesel combustion process and the high sulfur content in diesel fuel.

The cooling system used on a diesel engine must handle a greater heating load imposed on it by the oil cooler. Cooling system operation and components are very similar to their gasoline engine counterparts. Minor differences include a larger radiator and greater water pump capacity.

The coolant is comprised of ethylene glycol and water. The proper ratio of the two varies with the application and manufacturer.

CHAPTER 15 QUESTIONS

1. List the differences between a diesel and a gasoline engine lubricating system.
2. Why are oil coolers used and where are they located?
3. Why are piston cooling jets used and how do they operate?
4. Describe the S and C oil classifications.
5. How do soot and sulfur enter the lubricating system?
6. Describe the operation of the diesel cooling system.
7. Why is a high ratio of ethylene glycol to water used in some cases?

chapter 16
Intake, Exhaust, and Turbocharging Systems

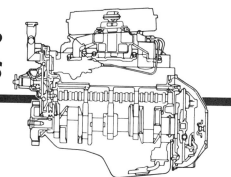

Objectives

In this chapter you will learn:

- Two special requirements of the diesel intake system
- Why the diesel exhaust system must be stronger
- The terms *naturally aspirated* and *supercharged*
- Why a naturally-aspirated engine does not achieve 100% volumetric efficiency
- How a turbocharger operates
- Five advantages of turbocharging a diesel engine
- How to service the intake, exhaust, and turbocharger systems

THE INTAKE SYSTEM

For the diesel engine to operate properly, it must be allowed to take in as much air as possible. The purpose of the intake system is to facilitate the intake of filtered air and effectively dampen engine noise. To accomplish these tasks, the diesel intake system is made relatively large and is usually equipped with a silencer (muffler), figure 16-1.

We have already seen that the diesel engine tries to draw in as much air as possible on every intake stroke since there is no throttle plate to limit incoming air. To allow the air to enter with few restrictions and to accommodate the diesel engine's need for large amounts of clean air, the air intake snorkel, air filter, and intake manifold are made as large as possible to satisfy the operating requirements.

The intake system has a fundamental influence on how well the diesel engine operates. If the intake system becomes restricted, there is not enough air to successfully burn all the fuel. This results in part of the injected fuel being heated only, changing to black smoke (soot). With the decrease in volumetric efficiency, the diesel engine reaches its smoke limit at lower RPM under high load conditions. Furthermore, since less fuel is being converted to mechanical energy, there is a loss of engine power. It is important that the intake system be kept free of debris and the air cleaner properly serviced.

The intake silencer is a muffling device usually mounted on the air cleaner intake. It acts much like a muffler on an exhaust system, except that it quiets the incoming air. The air cleaner canister also helps to reduce noise generated by the incoming air and the engine.

The intake manifold may be made of cast iron or aluminum. The intake manifold is larger to accommodate the extra air volume needed by the diesel. Often incorporated into the intake manifold is a crankcase ventilation device and an exhaust gas recirculation (EGR) valve. The crankcase ventilation device draws vapors from the crankcase into the intake manifold, relieving crankcase pressure. The EGR valve allows a portion of the exhaust gases to enter the intake air stream under certain operating conditions.

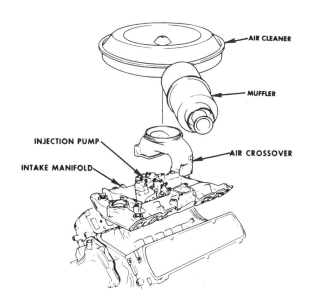

Figure 16-1 Air intake system (*Courtesy of General Motors Product Service Training*)

THE EXHAUST SYSTEM

The exhaust system on a diesel-powered car or truck is conventional in design with few modifications, figure 16-2. The muffler is tuned to the diesel's engine output, facilitating better scavenging of the exhaust gases. The exhaust manifold is made of cast iron and is usually larger to resist thermal shocks and vibrations. To help the exhaust system survive the vibrations created by the engine, special exhaust hanger brackets prevent the exhaust pipes from cracking and breaking. Note that there is no catalytic converter in the diesel exhaust system.

SERVICING THE INTAKE AND EXHAUST SYSTEM

The intake and exhaust manifolds should be thoroughly cleaned and inspected. Be careful not to scratch the gasket surfaces on aluminum sections. All gasket surfaces should be checked for warpage with a straightedge and feeler gauge. On some engines, there is no exhaust manifold gasket to take up slight imperfections, so the surface area must be free of defects and warpage.

Inspect the air filter canister and related hoses for signs of cracks or leaks. It is important that the air be filtered on the diesel engine and not find an alternate path. Unlike a gasoline engine, the diesel will run normally with an air leak on the intake manifold side without the owner knowing. Dirt will find its way in, prematurely wearing the engine.

TURBOCHARGERS

The diesel engines described so far are *naturally aspirated.* That means the engine receives a fresh charge of air using atmospheric pressure. A low-pressure area is created by the piston moving downward on the intake stroke, allowing air to be pushed in by atmospheric pres-

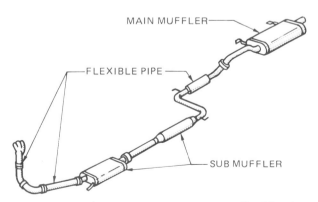

Figure 16-2 Diesel exhaust system uses flexible pipes, joints, and submufflers (resonators) on many models to reduce vibration and noise.

sure. Ideally, the engine should achieve 100% volumetric efficiency throughout its operating range. But in reality, a naturally-aspirated engine never achieves 100% volumetric efficiency. The point at which it comes closest to achieving 100% volumetric efficiency is the engine RPM at peak torque output. As volumetric efficiency decreases, so does engine torque. This is due to a number of factors including: restrictions in the intake and exhaust systems; the amount of time needed to fill the cylinder (time decreases as engine speed increases); increases in atmospheric temperature and humidity; the heating of intake air by the engine; increase in altitude; and type of combustion chamber.

Restrictions to air and exhaust gas flow are a result of traveling through: different-diameter pipes and hoses; sharp turns; the air filter and silencer; the muffler; and the intake and exhaust valves.

Obviously, many of these components are necessary for the engine to operate properly and durably, but they are restrictions that hinder air and exhaust gas flow.

As engine speed increases, there is less time to draw in a fresh charge of air since the intake valve is open the same duration (measured in degrees), at idle and top RPM. Because air does not start moving the instant the intake valve is open, there is short lag time in air movement. The lag time percentage of total air intake time increases as engine RPM increase, cutting down on air the cylinder receives.

As temperature, humidity, and altitude increase, there is a decrease in air density. The colder and drier the air, the greater the air density. Intake air temperature increases because of ambient conditions and engine heat. As humidity increases, air density decreases. Engineers have no control over altitude, humidity, and ambient temperature. They try to reduce heating of the intake air by the engine by ducting in air from outside the engine compartment, increasing volumetric efficiency.

Ideally, the engine should be 100% volumetric efficient, but on a naturally-aspirated engine, this is nearly impossible. Diesel engines are strongly influenced by volumetric efficiency. The greater the amount of air in the cylinder, the greater the amount of fuel that can be added, increasing the engine's power output. Engineers use external devices to increase the engine's volumetric efficiency and consequently its power output over a wider RPM range (approximately 40% increase in power). These devices *supercharge* the engine—they force a large quantity of air into the cylinder under pressure.

There are essentially two types of superchargers used on diesels. The Roots type is a positive displacement, engine-driven air pump. This pump is driven mechanically by the engine and displaces the same volume of air per revolution. The Roots type is found on large, two-stroke diesel engines and are absolutely essential for engine operation, figure 16-3. Another type of supercharger is the turbocharger, an exhaust-gas-driven, nonpositive displacement, centrifugal air pump, found on both heavy- and light-duty diesels. Currently it is the only type used on cars and light trucks. The turbocharger is driven by exhaust gas heat energy and velocity, which displace air by centrifugal force through the turbocharger's components. The volume of air displaced by the turbocharger per revolution varies.

The turbocharger is an integral part of the intake and exhaust systems, figure 16-4. A turbine wheel is placed in a path where exhaust gases exiting the exhaust manifold strike the turbine's blades, forcing it to spin very fast. The turbine is connected by a shaft (supported on an oil film) to a compressor wheel that rotates with the turbine. The compressor wheel, mounted in the air intake passage, draws the air from the air cleaner and accelerates it to a high velocity. The air travels to the diffuser section where the air's velocity is converted into pressure, often called the *boost pressure*, figure 16-5.

It is important that the turbine be located as close as possible to the exhaust manifold and withstand extremely high temperatures because the exhaust gases have the greatest heat energy and velocity here. More of the exhaust gases' heat energy and velocity will be transferred to the turbine, causing it to spin faster. If the turbine is located farther away from the exhaust manifold, the exhaust gases will cool and lose velocity; the turbine will not spin as fast, becoming less efficient. Special steel alloys are used to withstand the high temperatures and corrosive action of the exhaust gases.

The turbine and exhaust manifold are designed to complement each other. It is important that they match each other so exhaust passage will not be too restrictive and turbocharger *lag time* is minimized. Turbocharger lag is the time that turbocharger boost pressure is needed to when the boost pressure actually arrives at the combustion chamber. Ideally, this is instantaneous, but there is a time lapse once the operator steps on the throttle, increasing the amount of fuel; creating more and hotter exhaust gases, accelerating the turbine. Lag time can be reduced

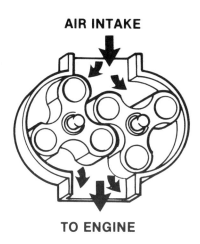

AIR INTAKE

TO ENGINE

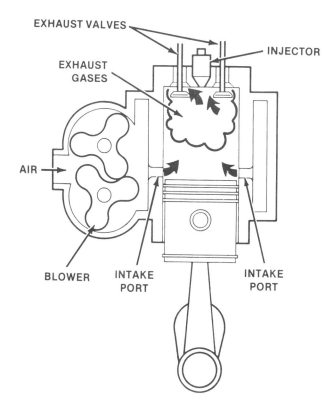

Figure 16-3 Two-stroke diesel engine needs an engine-driven air pump. *(Courtesy of Ford Motor Company)*

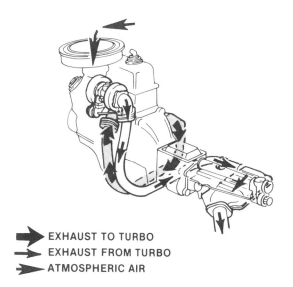

➤ EXHAUST TO TURBO
➤ EXHAUST FROM TURBO
➤ ATMOSPHERIC AIR

Figure 16-4 Exhaust gas flow through the turbocharger *(Courtesy of Ford Motor Company)*

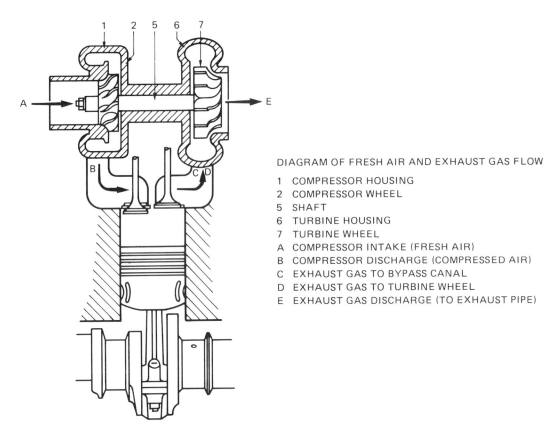

Figure 16-5 Cross section of the turbocharger *(Courtesy of Mercedes-Benz of North America Inc.)*

DIAGRAM OF FRESH AIR AND EXHAUST GAS FLOW

1 COMPRESSOR HOUSING
2 COMPRESSOR WHEEL
5 SHAFT
6 TURBINE HOUSING
7 TURBINE WHEEL
A COMPRESSOR INTAKE (FRESH AIR)
B COMPRESSOR DISCHARGE (COMPRESSED AIR)
C EXHAUST GAS TO BYPASS CANAL
D EXHAUST GAS TO TURBINE WHEEL
E EXHAUST GAS DISCHARGE (TO EXHAUST PIPE)

by narrowing the exhaust passage before the turbine, which speeds up the exhaust gases and the turbine at low engine RPM. But by compensating for lag time, the turbocharger may produce too much boost at higher engine RPM. This can be compensated for by a device called a wastegate that simply diverts some of the exhaust gases away from the turbine. The wastegate also acts as a boost control device. As boost pressure reaches a predetermined

level, the wastegate is fully open, preventing a condition known as *overboost*, which can seriously damage the engine, figure 16-6.

Turbocharger lag time can also be reduced by using smaller turbine and compressor wheels. Since they have less mass, they accelerate faster. Because they are smaller, however, they must spin very fast to achieve the proper volume of air flow. The turbine, compressor, and shaft

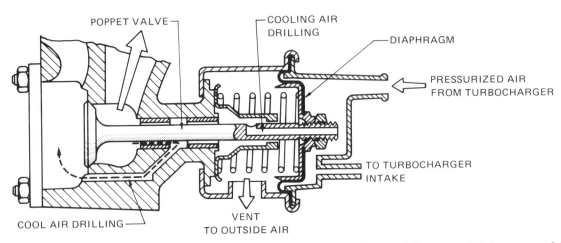

Figure 16-6 The wastegate controls exhaust gas flow through the turbocharger. *(Courtesy of Volkswagen of America Inc.)*

spin at speeds up to 130,000 RPM. This assembly must be balanced extremely well to prevent it from coming apart. No roller bearing could support the shaft at such tremendous speeds, so the shaft is supported by a film of oil in a manner similar to the crankshaft and connecting rod bearings. Oil under pressure is supplied by the engine oil pump, which not only lubricates the shaft and bearings but also helps keep them cool, figure 16-7.

There is a safety device often incorporated with turbocharger systems called a blow-off valve. The blow-off valve is located on the intake manifold, figure 16-8. Should boost pressure become too high, the valve vents the excess pressure to the atmosphere (air intake above turbocharger), preventing engine damage.

The turbocharger is well suited to the diesel engine and helps the engine overcome some of its traditional faults. Advantages of turbocharging a diesel are:

- higher engine torque and horsepower output
- improves power-to-weight ratio
- better fuel economy
- better combustion
- acts as an altitude compensator
- quieter running

The small, high-speed diesel develops peak torque at a low RPM, but torque drops off quickly as RPM increase. The turbocharger provides little assistance at low RPM but turbocharger output increases with engine RPM and

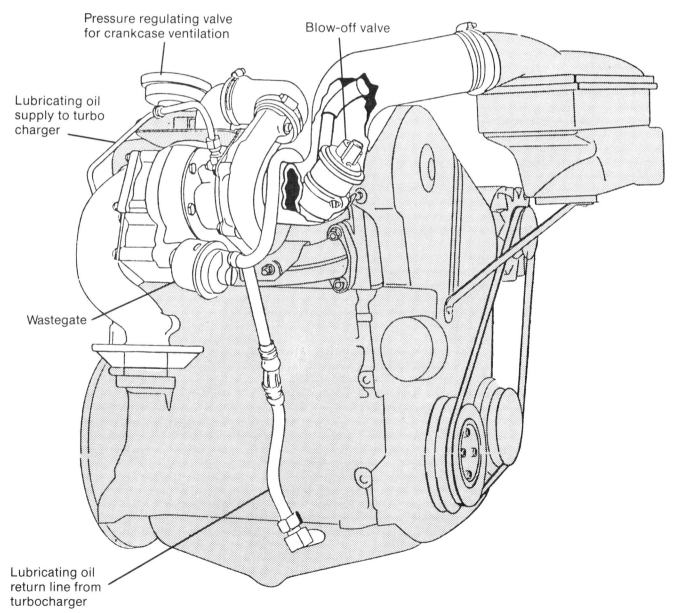

Pressure regulating valve for crankcase ventilation

Blow-off valve

Lubricating oil supply to turbo charger

Wastegate

Lubricating oil return line from turbocharger

Figure 16-7 Engine oil lubricates and cools the shaft and bearing. *(Courtesy of Volkswagen of America Inc.)*

Blow-off Valve

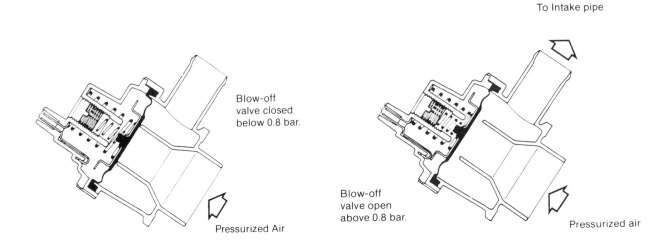

To Intake pipe

Blow-off
valve closed
below 0.8 bar.

Blow-off
valve open
above 0.8 bar.

Pressurized Air

Pressurized air

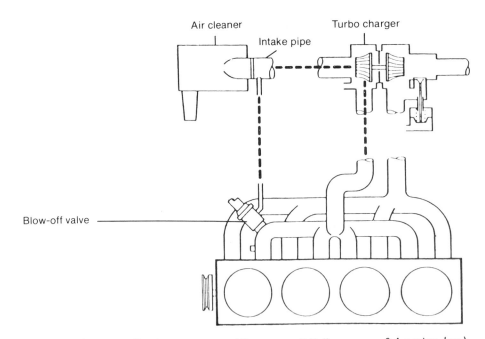

Air cleaner

Intake pipe

Turbo charger

Blow-off valve

Figure 16-8 Blow-off valve operation *(Courtesy of Volkswagen of America Inc.)*

load. Therefore, the diesel develops good power on its own at low RPM and the turbocharger provides the needed assistance as engine RPM and load increase. Boost pressure increases as load increases, because the more fuel put into the cylinder, the greater the exhaust gas volume and temperature. This also improves the engine's power-to-weight ratio, because torque and horsepower can be increased generally by 40% more than an aspirated engine of the same size, figure 16-9.

The higher horsepower and torque output allows the driver to select a higher gear ratio and shift less often. The result is better fuel economy, but this depends on how the vehicle is driven.

Using a turbocharger has other benefits as well. Since air is forced into the cylinder under high velocities, the air swirls and mixes easily with the injected fuel, providing good combustion. Under most conditions, the better combustion reduces exhaust emissions.

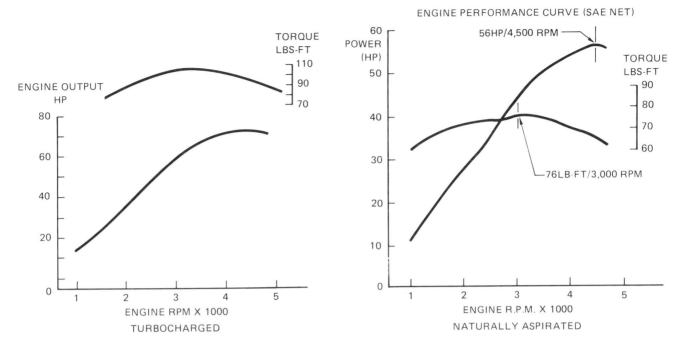

Figure 16-9 The turbocharged engine delivers higher horsepower and a high, sustained torque curve when compared to a naturally aspirated engine with the same displacement.

The turbocharger also acts as an altitude compensator. As altitude increases, the density of the air decreases. The turbocharger compensates for this loss by spinning faster. It spins faster because the decrease in air density produces a decrease in drag on the compressor wheel. By compensating for altitude, combustion is more thorough, and emissions are decreased.

The turbocharger quiets the engine by reducing intake and exhaust system resonance. The turbine and compressor wheels break up sound waves that normally travel through the open passages.

When compared to gasoline engines, diesel engines adapt easily to turbocharging. First, the diesel engine is already built rugged to handle high cylinder pressures. The addition of a turbocharger involves few modifications in diesels. Second, the diesel runs unthrottled and is relatively insensitive to air–fuel ratios, so the turbocharger can be set up to force into the cylinders as much air as the engine can safely handle.

There are minor drawbacks in using a turbocharger. First is the increased cost of adding a turbocharger unit and the modifications that must be made to the engine and fuel system. The engine may need modification to withstand the higher thermal loads. The fuel system needs modification to allow extra fuel into the cylinder when boost pressure is present. Second, though the turbocharger uses wasted heat energy, it is not a free ride entirely.

The turbocharger does impose a restriction on the engine in some modes of operation. Finally, particulate emission is higher than in a naturally-aspirated engine. This occurs on acceleration when fuel is injected into the cylinder, before the intake air is pressurized by the turbocharger.

Servicing the Turbocharger

The turbocharger, if defective, is replaced as a unit. However, it is important to determine the cause of failure. Owner abuse, such as shutting off the engine when extremely hot, causes the oil to change to carbon that seizes the shaft. Infrequent oil changes and dirt quickly ruin the turbocharger. Inspect the compressor and turbine wheels for signs of erosion caused by foreign material entering the intake and exhaust systems. Bent, broken, and missing blades are also cause for replacement.

The turbine shaft should spin freely without excessive roughness or freeplay. Bearing wear results from lack of oil, turbocharger overheating, and contaminated oil.

The turbocharger housing should be free of cracks and distortion. These may be caused by excessive vibration, improper mounting, thermal shocks, and physical abuse.

Whatever the cause of turbocharger failure, determine what led to its failure before installing a new unit.

When installing a new unit always be sure that the bearing and shaft are filled with oil.

SUMMARY

The intake system on a diesel must dampen engine noise and allow the engine to draw in as much filtered air as possible. The air passages and filter are larger to accommodate the extra volume of air, and a silencer is used to control noise.

The exhaust system is much like a conventional exhaust system used on gasoline-powered vehicles. Differences include larger-diameter pipes, heavier exhaust manifolds to withstand thermal shock, and heavier-duty brackets and clamps to secure the system from engine vibrations.

On a naturally-aspirated engine, atmospheric pressure forces the air into the engine's cylinders. A naturally-aspirated engine never achieves 100% volumetric efficiency because of ambient conditions, restrictions in the intake and exhaust systems, and lack of time to fill the cylinder.

To increase volumetric efficiency, superchargers are added. The type used on small, high-speed diesels is the turbocharger. The turbocharger is an exhaust-gas-driven, nonpositive displacement centrifugal air pump. Exhaust gases exiting the exhaust manifold strike the turbine blades forcing them to revolve at high speed. The turbine is connected by a shaft to a compressor wheel that accelerates the intake air. The high-velocity air enters a diffuser where the energy from the air velocity is converted to pressure called the boost pressure.

To increase exhaust gas flow and prevent boost pressure from becoming too high, a wastegate is used. The wastegate opens an alternate path around the turbine for the exhaust gases.

In the event of malfunction, a blow-off valve is used to vent excessive boost pressure to the atmosphere, preventing engine damage.

There are several advantages to turbocharging a diesel engine. It increases horsepower and torque, improves power-to-weight ratio, promotes better combustion, lowers exhaust emissions, acts as an altitude compensator, and helps quiet the engine.

Diesel engines adapt to turbocharging more easily than gasoline engines. Engine modifications are few and the diesel is insensitive to air-fuel ratios.

The disadvantages of using a turbocharger are increased costs; modifications must still be made to handle the high thermal loads; it is a restriction in some modes of operation; the fuel system must be modified; and particulate emissions are higher, especially on acceleration.

CHAPTER 16 QUESTIONS

1. Explain two special requirements of the diesel intake system.
2. Explain why air leak on the intake side of a diesel must be repaired.
3. Why must the diesel exhaust system be made stronger?
4. Why does a naturally-aspirated engine not achieve 100% volumetric efficiency?
5. Explain how a turbocharger operates.
6. List and explain five advantages of turbocharging a diesel engine.
7. List the disadvantages of using a turbocharger.

chapter 17

Diesel Engine
Assembly and Installation

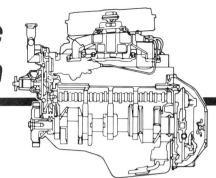

Objectives

In this chapter you will learn:
- **How to correctly assemble a diesel engine**
- **Why a valve lifter bleed-down procedure may be needed**
- **How to correctly install the engine**

DIESEL ENGINE ASSEMBLY

The procedure for assembling a diesel engine is very similar to procedures used for gasoline engines. However, there are more precautions that must be adhered to and some areas of assembly may be subtly different. The following is a general guide.

This task requires the following:

- appropriate hand tools
- proper gasket set
- proper chemical sealers (if needed)
- manufacturer's service manual

To avoid contamination be sure the work area is clean. Use of chemical sealers is now quite common. Be sure the proper chemical sealers are on hand to save time. Be sure all threaded surfaces are clean by running a tap down each hole. Lubricate all moving and sliding surfaces (bearings, for example) with the correct engine oil.

1. Install the crankshaft. Be sure the bearings are installed correctly. Hand tighten the bearing cap bolts. Align the thrust bearing, figure 17-1. Torque the bearing cap bolts to manufacturer's specifications. Check the drag by rotating the crankshaft by hand as each bearing cap is tightened. Check crankshaft end play for the proper clearance.
2. Install the pistons and connecting rods. Install the piston rings according to the manufacturer, figure 17-2. Cover the connecting rod bolts with a hose to prevent damage, figure 17-3.

CAUTION: It is absolutely essential that the piston be installed in the proper cylinder bore facing the proper direction, figure 17-4. Failure to do so can result in severe engine damage.

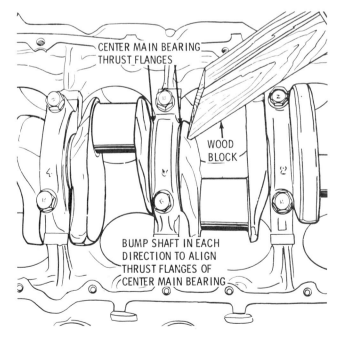

Figure 17-1 Aligning the crankshaft thrust bearing (*Courtesy of Oldsmobile Division*)

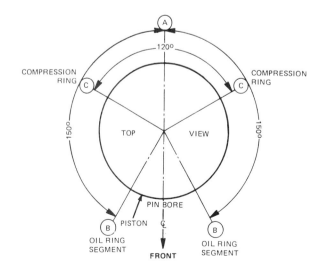

Figure 17-2 Piston ring alignment (*Courtesy of Ford Motor Company*)

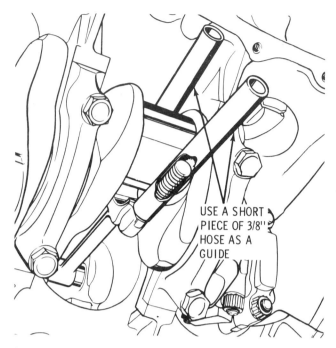

Figure 17-3 Protecting the rod-bearing journal during installation (*Courtesy of Oldsmobile Division*)

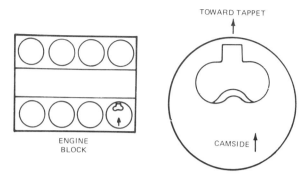

Figure 17-4 The pistons must be installed properly to prevent damage. (*Courtesy of Ford Motor Company*)

Compress the piston rings and gently tap the piston into the cylinder. Tighten the connecting rod caps to manufacturer's specifications. Rotate the crankshaft by hand after each connecting rod cap is tightened to check for excessive drag. Check connecting rod side clearance, figure 17-5.

3. Install the oil pump and pick-up screen.
4. Install the camshaft if it is located in the block. Next, install the timing gears or chain.

CAUTION: Be sure all the timing marks line up, figure 17-6. If they are not timed properly, serious engine damage could result.

---

5. Install the front cover or timing gear case.
6. Install the oil pan and gasket.
7. Install the flywheel.
8. Install the oil cooler if located on the block.
9. Install the head gasket(s). Check for proper alignment between all passages. Be sure the gasket is mounted with the correct surface facing the proper direction.

CAUTION: Use no sealer or gasket cement unless specified by the manufacturer.

---

10. Install the cylinder head(s), figure 17.7.

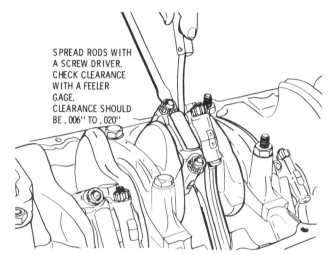

Figure 17-5 Checking connecting rod clearance (*Courtesy of Oldsmobile Division*)

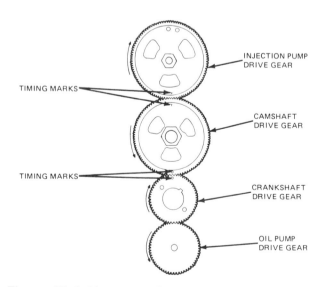

Figure 17-6 Alignment of timing marks is critical to prevent engine damage. (*Courtesy of Ford Motor Company LTD*)

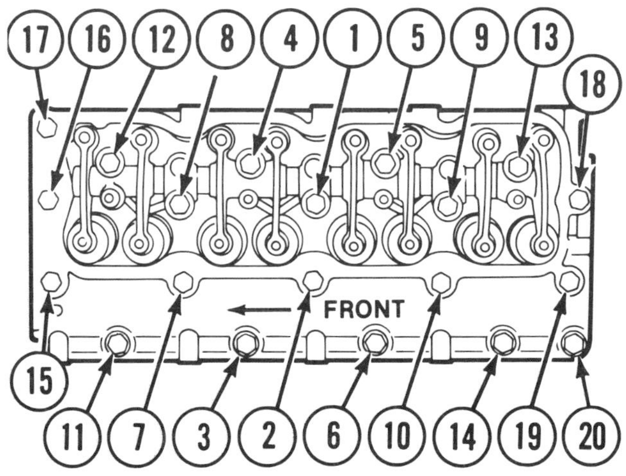

Figure 17-7 Typical cylinder head bolt-tightening sequence (*Courtesy of Ford Motor Company*)

CAUTION: Be sure the bolts are clean. Dip them in oil if the manufacturer specifies. Torque the cylinder head bolts in the proper pattern and sequence, according to manufacturer's directions.

11. Install the camshaft if located in cylinder head and timing drive.
12. Install the glow plugs with a torque wrench. Use of anti-seize compounds may be required.
13. Install the injection nozzles. Always use new gaskets and seals when installing injection nozzles.
14. Install the lifters in the proper bores. With roller lifters be certain the roller rotates with the camshaft. Install the retainers and guides. Rotate the

engine to be sure the lifters move freely in their bores.

15. Install the pushrods.

CAUTION: Be sure each pushrod has the hardened tip in the correct position. With roller lifters, the hardened tip is at the top.

16. Install the rocker arms and hold-downs.

CAUTION: Rocker arms and hold-downs must be installed according to the manufacturer's directions, figure 9-11.

With hydraulic lifters, a valve lifter bleed-down procedure may be required. This procedure is required whenever the rocker arms are loosened or removed. The lifter may become extended in length when force is removed from it or when it has just been rebuilt. Serious damage to the engine can result if it is not bled down (the valve can contact the piston). Generally, the valve lifter bleed-down procedure entails rotating the crankshaft slowly to a specified position, and tightening specific rocker arms to the specified torque. The crankshaft position is held a certain length of time to allow the valve lifters to bleed down. This procedure may require repeating, with the crankshaft in different positions to cover any remaining valve lifters. Consult the service manual for the specific procedure.

17. Install the intake manifold, figure 17-8.
18. Install the valve cover(s).

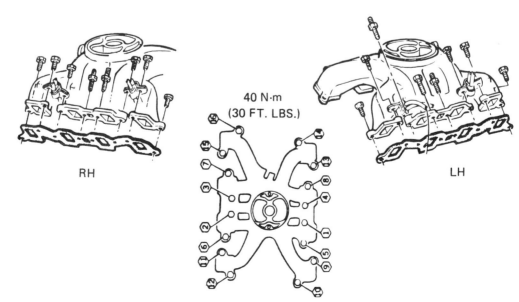

40 N·m
(30 FT. LBS.)

RH

LH

Figure 17-8 Intake manifold bolt-tightening sequence and location of special bolts (*Courtesy of Chevrolet Division*)

19. Install the injection pump. Be sure the injection pump is timed correctly with the engine.
20. Install the crankshaft dampener.
21. Install the water pump.
22. Install the pulleys.
23. Install the fuel pump, lines, and filters.

Make an inspection of the engine before it goes into the vehicle. Have all openings covered to prevent debris from entering the engine.

ENGINE INSTALLATION

1. Secure the lifting device to the appropriate points on the engine. Disconnect the engine from the stand.
2. Lower the engine into the vehicle. Connect the engine to the transmission and engine mounts.
3. Connect the exhaust pipe(s).
4. Connect the throttle linkage.
5. Connect the electrical harness and starter cables.
6. Connect the heater hoses, vacuum lines, power steering hoses, fuel lines, and air-conditioning compressor.
7. Install the radiator and support brackets.
8. Connect the radiator hoses, engine oil and transmission oil lines.
9. Fill the crankshaft, radiator, and power steering pump with the appropriate fluids.
10. Connect the battery cable(s) and ground strap.
11. It is highly recommended that the lubrication system be primed to prevent premature engine wear and failure. Follow the manufacturer's directions on priming the system. If no primer is available, there is an alternate method. Disconnect the injection pump fuel shut-off. Disconnect the glow plug harness and remove all the glow plugs. Crank the engine, but for no more than 20 seconds at a time to prevent the starter motor from overheating. The oil light should go out. If it does not, repeat the procedure in a few minutes.

12. Start the engine. It may start hard at first because of air in the fuel lines.

CAUTION: Do not use any starting aids or gasoline. This may seriously damage the engine.

---

13. Adjust the idle speed and look for any signs of oil, coolant, compression, and fuel leakage.
14. Install and align the hood.
15. Road test the vehicle to be sure the engine has good power and the transmission shifts properly.
16. Conduct a final inspection under the hood to be sure everything is in order. Clean the vehicle when done.

SUMMARY

Care and caution are needed for a successful engine overhaul. Small details that may be neglected in a gasoline engine cannot be neglected in a diesel engine. Important items to remember are:

● Lubricate all moving and sliding surfaces.
● Be sure the component is in the correct position and place. For example, a piston that is match fitted to a particular cylinder bore should be installed in that bore.
● Use the correct chemical sealers. Do not use any sealer on the cylinder head gasket unless specified.
● The valve lifters may require bleeding down to prevent contact between the valves and pistons.
● All bolts must be torqued to the specified value in the proper pattern.
● All openings are either guarded or capped to prevent personal injury and damage to the engine.

Once the engine is installed in the vehicle and running, be sure the customer will be satisfied with the results. A vehicle road test may reveal other problems or adjustments that must be corrected. Finally, be sure the vehicle is clean when the owner recieves it.

CHAPTER 17 QUESTIONS

1. On a diesel engine, alignment of timing marks is critical. Why?
2. Why must hydraulic valve lifters be bled?
3. Describe the hydraulic valve lifter bleed-down procedure.
4. Why should the lubrication system be primed?

chapter 18
Maintenance and Diagnosis

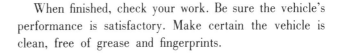

Objectives

In this chapter you will learn:
- **The basic maintenance procedures**
- **The procedure for diagnosing a no-starting complaint**
- **The procedure for diagnosing a low-power complaint**
- **The procedure for locating a missing cylinder**
- **The glow plug resistance test procedure**
- **The problems that could cause black smoke**
- **The problems that could cause white smoke**
- **The problems that could cause blue smoke**

DIESEL ENGINE MAINTENANCE

Like any other engine, the diesel engine requires maintenance at specified intervals. Typical maintenance items are oil and filter service; valve adjustment; compression check; air filter replacement; fuel filter service; crankcase ventilation system service; engine idle speed check; fuel injection pump timing check; EGR inspection and cleaning; EPR valve inspection, and fuel nozzle check.

When these items are serviced depends on the manufacturer's service schedule, and driving conditions. Manufacturer's service schedules will vary. Driving conditions influence when the vehicle should be serviced, figure 18-1. Short trips in which the engine rarely gets warmed up, or dusty conditions should be considered as severe operating conditions requiring frequent service. Finally, the owner must be aware of conditions the vehicle is operating in and bring the vehicle in at the appropriate times. Be sure the owner knows that the diesel-powered vehicle has different needs from the gasoline-powered vehicle.

Performing the necessary services according to the manufacturer's service manual requires the following:

- appropriate hand tools
- appropriate fluids and filters
- tachometer
- timing meter
- nozzle tester

Consult the necessary chapters in this book and the manufacturer's service manual when performing the specified maintenance. Idle speed, timing, and other services listed should be adjusted according to the underhood engine specifications label.

When finished, check your work. Be sure the vehicle's performance is satisfactory. Make certain the vehicle is clean, free of grease and fingerprints.

DIAGNOSIS

When diagnosing a system, there should be a format to follow that leads the technician to the problem safely, logically, and quickly. What follows is one method of performing a diagnosis. But no matter what format is chosen the technician must care about the work to achieve positive results.

Diesel engine diagnosis begins with listening to the owner. First, ask the owner to describe the complaint. Note when the symptoms occur and where they occur. Do not be fooled by the owner's conclusions about what is wrong or what should be done.

Second, confirm that the complaint exists and note any other symptoms not mentioned by the owner. A road test may be needed, perhaps with the owner to stimulate the condition that causes the problem. It is here that operator error can be detected and corrected. For example, the driver complains of hard starting. The problem may simply be that the driver is not waiting for the glow plug light to go out before starting. Explain to the owner in a courteous manner the correct starting procedure and why the procedure should be followed.

Third, list in order of probability the likely causes of the complaint. You must know what systems are in operation, how they operate, and how they affect each other. Consulting troubleshooting guides at this point is useful.

Fourth, isolate the problem. You must test and check the probable systems for the cause of the problem. How-

SCHEDULED MAINTENANCE

SERVICE INTERVALS													
Perform at the months (*) or distances shown,	Miles	7500	15,000	22,500	30,000	37,500	45,000	52,500	60,000	67,500	75,000	82,500	90,000
whichever comes first	Kilometers	12 070	24 140	36 210	48 280	60 350	72 420	84 000	96 000	108 000	120 000	132 000	144 000
Change Engine Oil*		X	X	X	X	X	X	X	X	X	X	X	X
Replace Engine Oil Filter	Main*	X	X	X	X	X	X	X	X	X	X	X	X
	Bypass		X		X		X		X		X		X
Check Coolant Condition and Protection					— ANNUALLY —								
Check Cooling System / Hoses / Clamps					— ANNUALLY —								
Replace Coolant					X				X				X
Check Drive Belts on Engine and Accessories		X			X				X				X
Adjust Engine Valve Clearance			X		X	X							
Replace Air Cleaner Element***						X			X				X
Drain Water from Fuel Conditioner		X	X	X	X	X	X						
Check Injection Pump Timing			X										

*Change engine oil and main filter every 7500 mi (12 070 km) or 7.5 months, whichever comes first.
**Replace camshaft and fuel injection pump drive belts every 100,000 mi (160 000 km).
***Replace air cleaner element more often if operating under severe dust conditions.

SEVERE SERVICE
If your driving habits include frequent short trips of 10 mi (16 km) or less **when the temperature remains below 10°F (−12°C) for 60 days or more**; sustained high-speed driving **during hot weather** (90°F, 32°C); towing a trailer for long distances, driving in severe dust conditions; extensive idling, such as police, taxi, or door-to-door delivery use; or extended periods of snow plowing, the following severe maintenance intervals apply:

Engine Oil and
Main Oil Filter
— Change every three months or 3000 miles (4 828 km), whichever comes first.

Bypass Oil Filter
— Change every six months or 6000 miles (9 656 km), whichever comes first.

Air Cleaner Filter
— If operating in severe dust conditions, replace more often than regular intervals, as determined by local procedures.

Figure 18-1 Typical service schedule for a diesel-powered car (*Courtesy of Ford Motor Company*)

ever, do not overlook the obvious, simple items. For example, the owner complains of excessive black smoke and loss of power. True, the problem may be in the nozzles or injection pump timing, but it makes more sense to check the air filter first to see if it is clogged with debris.

Fifth, repair the problem. Once the necessary repairs and adjustments are completed, check to see if the problem is corrected.

1. ABNORMAL CONDITIONS, CAUSES, AND CORRECTIONS

Engine Will Not Crank

1. Loose or corroded battery cables. Check all cable connections.

2. Discharged battery(ies). Recharge the battery(ies) and check the charging system output.
3. Starter inoperative. Check voltage and amperage to starter.

Engine Cranks Slowly—Will Not Start (Minimum cranking speed—100 RPM COLD, 200 RPM HOT)

1. Loose or corroded battery cables. Check all cable connections.
2. Battery(ies) undercharged. Recharge the battery(ies) and check charging system output.
3. Wrong engine oil. Drain and refill with recommended oil.

Engine Cranks Normally—Will Not Start

1. Incorrect starting procedure. Use the recommended procedure.
2. Glow plug system inoperative. Refer to "Glow Plug Resistance Procedure."
3. No fuel to nozzle. Loosen the nozzle line at the nozzle. Use care to direct fuel away from sources of ingition. Wipe the connection dry. Crank for 5 seconds. Fuel should flow from the nozzle line (go to step 9). Tighten the connection.
4. Fuel shut-off solenoid inoperative. Check the fuel shut-off solenoid operation by repeatedly turning the ignition switch on/off. There should be a clicking noise indicating the solenoid is working. If no noise is heard, connect a voltmeter to the wire at the fuel shut-off solenoid and ground. Minimum voltage is 9 volts. If under 9 volts, check the circuit. If voltage is within specifications, replace the fuel shut-off solenoid.
5. No fuel to injection pump (if equipped with external supply pump). Loosen the fuel line coming out of the filter. Crank the engine. Fuel should come out of the filter (direct fuel away from sources of ignition). If fuel sprays out, go to step 8.
6. Restricted fuel filter. Loosen the fuel line going to the filter. If fuel sprays from the fitting while cranking, the filter is plugged and must be replaced.
7. Fuel supply pump inoperative. Remove the inlet hose to the fuel pump. Connect a hose to the fuel pump from a separate container that contains fuel. Loosen the line going to the filter. If fuel does not spray from the fitting, replace the pump. Use care to direct fuel away from sources of ignition.
8. Restricted fuel line and fuel tank strainer. Inspect the fuel line for dents and kinks. If no restriction is found, remove the tank and inspect the strainer.
9. Plugged return fuel line. Disconnect the return line at the injection pump and connect a hose from the injection pump to a container. Crank the engine. If it starts, correct the restriction in the return fuel circuit.
10. Incorrect or contaiminated fuel. Flush the fuel system and install correct fuel.
11. Pump timing incorrect. Check and adjust timing.
12. Faulty nozzle(s). Remove the nozzle(s) and the test with the nozzle tester.
13. Low compression. Check cylinder compression.

Engine Starts but Will Not Continue to Run at Idle

1. Slow-idle speed incorrect. Adjust idle speed to specifications.
2. Fast-idle solenoid inoperative. With the engine cold, the solenoid should move to hold the injection pump lever in the fast-idle position. If the solenoid does not move, check the fast-idle circuit.
3. Restricted fuel return system. Disconnect the return line at the injection pump and connect a hose from the injection pump to a container. Crank the engine and allow it to idle. If the engine idles normally, correct the restriction in the fuel return line circuit. If the engine does not idle normally, remove the return line check valve (DB2 pumps only) and be sure it is not plugged.
4. Glow plugs turn off too soon. Refer to "Glow Plug Resistance Procedure."
5. Pump timing incorrect. Check and adjust timing to specifications.
6. Limited fuel to injection pump. Check the fuel filter, fuel lines, and fuel supply pump. Replace or repair as necessary.
7. Incorrect or contaiminated fuel. Flush the fuel system and install correct fuel.
8. Low compression. Check cylinder compression.
9. Fuel shut-off solenoid closes in the run position. Ignition switch out of adjustment.
10. Injection pump malfunction. Remove the injection pump for repair.

Excessive Surge at Light Throttle, under Load

1. If equipped with a torque converter clutch (TCC), the TCC is engaging too soon. Refer to an automatic transmission book on TCC diagnosis.
2. Injection pump timing retarded. Check and adjust timing to specifications.
3. Restricted fuel filter. Check fuel pump pressure at inlet and outlet sides of the fuel filter. Replace the filter, if necessary.
4. Injection pump housing pressure too high. Refer to HPCA diagnosis.

Engine Starts and Idles Rough with Excess Noise and/or Smoke, but Clears up after Warmup

1. Incorrect starting procedure. Advise the owner on the correct procedure.
2. Injection pump timing incorrect. Check and adjust timing to specifications.
3. Air in system. Install a section of clear plastic tubing on the fuel return line and start the engine. Evidence of bubbles during cranking or running indicates an air leak on the suction side of the fuel system.

4. Inoperative glow plugs. Locate and replace the defective glow plug(s).
5. Nozzle(s) malfunction. Remove, test, clean, or replace the nozzle(s).
6. HPCA inoperative. Check the HPCA circuit.

Engine Misfires above Idle, but Idles Correctly

1. Restricted fuel filter. Check and replace the filter.
2. Incorrect injection pump timing. Check and adjust timing.
3. Incorrect or contaminated fuel. Flush the fuel system and fill it with correct fuel.

Rough Engine Idle

1. Incorrect idle speed. Set idle to manufacturer's specifications.
2. Air in system. Check for the presence of air by installing clear plastic tubing in the return fuel line. Repair the leak and purge the air from the fuel system.
3. Incorrect injection pump timing. Check and adjust timing to specification.
4. Leaking nozzle lines. Visually inspect the lines at both pump and nozzles. Tighten or replace the lines.
5. Low fuel pressure between filter and injection pump. Check fuel pressure. Repair the restriction.
6. Faulty nozzle(s). Determine which cylinder(s) may have the faulty injectors. Remove, test, clean or replace the nozzle(s).
7. Low cylinder compression. Perform a compression test to determine cylinder pressure.
8. If objectionable idle quality still exists, go to glow plug resistance check.

Engine Will Not Return to Idle

1. External linkage binding or misadjusted. Free the linkage. Adjust or replace as required.
2. Fast-idle system malfunction. Check fast-idle adjustment and circuit.
3. Internal injection pump malfunction. Remove the injection pump for repair.

Fuel Leaks on the Ground—No Engine Malfunction

1. Loose or broken fuel line, connection, or both. Examine the fuel system, repair as needed.
2. Injection pump seal(s). Remove the injection for repair and replace the seal(s).

Noticeable Loss of Power

1. Restricted air intake. Check the air cleaner element and passages.
2. Injection pump timing not set to specifications. Check and adjust pump timing.
3. EGR malfunction (if equipped). Refer to Chapter 11, "EGR/EPR System Service."
4. Restricted or damaged exhaust system. Check the system and replace as necessary.
5. Restricted fuel filter. Check and replace the filter.
6. Plugged fuel tank vacuum vent in fuel cap. Remove the fuel cap. If loud hissing sound is heard, the vacuum vent is plugged. Replace the cap. (Slight hissing sound is normal.)
7. Restricted fuel supply from fuel tank to injection pump. Examine the fuel supply system and determine the cause of restriction. Repair as required.
8. Restricted fuel tank strainer. Remove the fuel tank and check the strainer.
9. Restricted return line fuel circuit. Examine and check the return fuel circuit for restrictions and correct as required.
10. Incorrect or contaminated fuel. Flush the fuel system and install correct fuel.
11. External compression leaks. Check for compression leaks at all nozzles and glow plugs. If the leak is found, tighten the nozzle(s) or glow plugs.
12. Faulty nozzles. Remove the nozzles; check and clean them, and replace if necessary.
13. Low cylinder compression. Check compression to determine the cause.

Noise—Rap from One or More Cylinders (Sounds Like Rod Bearing Knock)

1. Nozzle(s) sticking open or very low nozzle opening pressure. Remove the nozzle from the noisy cylinder(s). Test, clean, or replace the nozzle.
2. Mechanical problem. Refer to "Engine Mechanical Diagnosis."
3. Piston hitting cylinder head. Replace the malfunctioning parts.

Noise—Objectionable Overall Combustion Noise over Normal Noise Level with Excessive Black Smoke

1. Timing not to set to specification. Check and adjust timing to specifications.
2. EGR malfunction. Refer to Chapter 11, "EGR/EPR System Service."
3. Injection pump housing pressure out of specifications.

Refer to Chapter 6, "Injection Pump Housing Pressure Diagnosis."

4. Injection pump internal problem. Remove the injection pump for repair.

Engine Noise—Internal or External

1. Engine fuel pump, generator, water pump, valve train, vacuum, pump, bearings, etc. Repair or replace as necessary. If the noise is internal, see "Noise—Rap from One or More Cylinders" and "Engine Starts and Idles Rough with Excessive Noise and Smoke."

Engine Overheats

1. Coolant system leak, oil cooler system leak, or coolant recovery system not operating. Check for leaks and correct as required. Check the coolant recovery jar, hose, and radiator cap.
2. Belt slipping or damaged. Replace or adjust as required.
3. Thermostat stuck closed. Check and replace if required.
4. Head gasket leaking. Check and repair as required.

Instrument Panel Oil Warning Lamp "on" at Idle

1. Oil cooler or cooler line restricted. Remove the restrictions in cooler or cooler line.
2. Oil pump pressure low. Refer to "Low Oil Pressure." Engine Will Not Shut Off with Key.

Note: With engine at idle, pinch the fuel return line at the flexible hose to shut off engine.

1. Injection pump fuel solenoid (or controller if equipped) does not return fuel valve to off position. Refer to the manufacturer's section on diagnosis.
2. If equipped with vacuum shut-off, check all lines.

2. ENGINE MECHANICAL DIAGNOSIS
Excessive Oil Loss

1. External oil leaks. Tighten bolts or replace gaskets and seals as necessary.
2. Improper reading of dipstick. Check the oil with the car on a level surface and allow adequate draindown time.
3. Improper oil viscosity. Use recommended SAE viscosity for prevailing temperatures.
4. Continuous high-speed driving or severe usage such as trailer hauling. Both of these factors normally cause decreased oil mileage.

5. PCV (CDR) system malfunctioning. Service as necessary.
6. Valve guides or valve stem seals worn, or seals omitted. Ream the guides and install oversize service valves or new valve stem seals.
7. Piston rings not seated, broken or worn. Allow adequate time for the rings to seat. Replace broken or worn rings as necessary.
8. Piston improperly installed or misfitted. Replace the piston or repair as necessary.

Low Oil Pressure

1. Slow idle speed. Set idle speed to specifications. Replace with proper switch.
2. Incorrect or malfunctioning oil pressure gauge. Replace with proper gauge.
3. Improper oil viscosity or diluted oil. Install oil of proper viscosity for expected temperature. Install new oil if diluted with moisture or unburned fuel mixtures.
4. Oil pump worn or dirty. Clean the pump and replace worn parts as necessary.
5. Plugged oil filter. Replace the filter and oil.
6. Oil pickup screen loose or plugged. Clean or replace the screen as necessary.
7. Hole in oil pickup tube. Replace the tube.
8. Excessive bearing. Replace as necessary.
9. Cracked, porous, or plugged oil galleys. Repair or replace the block.
10. Galley plugs missing or misinstalled. Install plugs or repair as necessary.
11. Poor seal at timing cover gasket. Replace the gasket.

Valve Train Noise

1. Low oil pressure. Repair as necessary. (See "Low Oil Pressure.")
2. Loose rocker arm attachments. Inspect and repair as necessary.
3. Worn rocker arm, pushrod, or both. Replace as necessary.
4. Broken valve spring. Replace the spring.
5. Sticking valves. Free the valves.
6. Lifters worn, dirty, or defective. Clean, inspect, test, and replace as necessary.
7. Camshaft worn or poor machining. Replace the camshaft.
8. Worn valve guides. Repair as necessary.

3. ENGINE KNOCK DIAGNOSIS

Engine Knocks Cold and Continues for 2 to 3 Minutes, Knock Increases with Torque

1. Flywheel contacting splash shield. Reposition the splash shield.
2. Loose or broken balancer or drive pulleys. Tighten or replace as necessary.
3. Excessive piston-to-bore clearance. Replace the piston.
4. Bent connecting rob. Replace the bent connecting rod.

Engine Has Heavy Knock when Hot with Torque Applied

1. Broken balancer or pulley hub. Replace parts as necessary.
2. Loose torque converter. Tighten the bolts.
3. Accessory belts too tight or nicked. Replace or tension to specifications as necessary.
4. Exhaust system grounded. Reposition as necessary.
5. Flywheel cracked. Replace the flywheel.
6. Excessive main bearing clearance. Replace as necessary.
7. Excessive rod bearing clearance. Replace as necessary.

Engine Knocks on Initial Startup but Only for a Few Seconds

1. Fuel supply pump. Replace the pump.
2. Improper oil viscosity. Install proper oil viscosity for expected temperatures.
3. Hydraulic lifter bleed-down. Clean, test, and replace as necessary.
4. Excessive crankshaft end clearance. Replace the crankshaft thrust bearing.
5. Excessive front main bearing clearance. Replace worn parts.

Engine Knocks at Idle when Hot

1. Loose or worn drive belts. Adjust or replace as necessary.
2. Compressor or generator bearing. Replace as necessary.
3. Fuel supply pump. Replace the pump.
4. Valve train. Replace parts as necessary.
5. Improper oil viscosity. Install proper-viscosity oil for expected temperature.
6. Excessive piston pin clearance. Replace the piston, piston pin, and connecting rod if necessary.
7. Connecting rod alignment. Check and replace rods as necessary.
8. Insufficient piston-to-bore clearance. Hone and fit a new piston.
9. Loose crankshaft balancer. Torque or replace worn parts.

Missing Cylinder Diagnosis

Two primary methods are used for diagnosing a missing (skipping) cylinder. The first is similar to that used on gasoline engines for measuring individual cylinder output. This task requires the following:

- Appropriate hand tools
- Tachometer
- Manufacturer's service manual

Diagnose a missing cylinder by following these procedures:

1. Connect the tachometer and adjust engine RPM to specified speed.
2. Loosen the nozzle line and note RPM drop.

CAUTION: Direct fuel away from sources of combustion.

---

3. Repeat the procedure for each cylinder. The cylinder producing little or no RPM drop is the missing cylinder. Generally, the drop in RPM between cylinders should be no more than 10 RPM for a smooth-running engine. If not, refer to "Noticeable Loss of Power."

Glow Plug Resistance Procedure

1. Use a high impedance digital multimeter for resistance measurements.
2. Select an appropriate scale.
3. Start the engine, turn on the heater, and allow the engine to warm up. REMOVE all the feed wires from the glow plugs.
4. Disconnect the generator (alternator) two-lead connector.
5. Using a tachometer, adjust engine speed by turning the idle speed screw on the side of the injection pump to the worst engine idle roughness; but do not exceed 900 RPM.
6. Allow the engine to run at worst idle speed for at least 1 minute. The thermostat must be open and the upper radiator hose hot.
7. Attach an alligator clip to the black test lead of the multimeter.

CAUTION: This clip must be grounded to a specified point (such as the fast-idle solenoid).

It must remain grounded to this point until all tests are completed.

8. On a separate sheet of plain writing paper write down the engine-firing order.

9. With the engine still idling, probe each glow plug terminal and record the resistance values on each cylinder in firing sequence. Most readings will be from 1.8 to 3.4 ohms. If these readings are not obtained, turn the engine "off" for several minutes and recheck the glow plugs. The resistance should be .7 or .8 ohm. If this reading is not obtained, check the meter for correct settings, check for low or incorrect battery in the meter, and check the meter ground wire to the engine.

10. The resistance values are dependent on the temperature in each cylinder, and therefore indicate the output of each cylinder. The higher the temperature, the greater the resistance.

11. If the ohm reading on any cylinder is about 1.2 or 1.3 ohms, check if there is a mechanical engine problem. Make a compression check of the low-reading cylinder and the cylinders which fire before and after the low cylinder reading. Correct the cause of the low compression before proceeding to the fuel system.

12. Examine the results of all cylinder glow plug resistance readings, looking for differences between cylinders. Normally, rough engines will have a difference of .3 ohm or more between cylinders in firing order. It will be necessary to raise or lower the reading on one or more of these cylinders by selection of injection nozzles.

13. Remove the injection nozzles from the cylinders in which the ohm reading needs to be raised or lowered. Determine the pop-off (opening) pressure of the nozzles, and check the nozzle for leakage and spray pattern. (Refer to Chapter 9, "Nozzle Testing."

a. Install injection nozzles with a higher (pop-off) pressure to lower the ohm reading, and injection nozzles with lower pop-off pressure to raise an ohm reading. Normally, a change of about 30 psi in pressure will change the reading by .1 ohm. Injection nozzles normally drop off in pop-off pressure. Use nozzles from parts stock or a new car. Use broken-in nozzles on a car with 1500 or more miles, if possible.

b. Whenever an injection nozzle is cleaned or replaced, crank the engine and watch for air bubbles at the injection nozzle inlet before installing the injection line. If bubbles are present, clean or replace the injection nozzle.

c. Install the injection line, restart the engine, and check the idle quality. If idle is still not acceptable, recheck the glow plug resistance of each cylinder in firing-order sequence. Record readings.

d. Examine all glow plug resistance readings, looking for differences of .3 ohm or more between cylinders. It is necessary to raise or lower the reading on one or more of these cylinders as previously done.

e. After making additional injection nozzle changes, check idle quality again. Normally, after completing two series of resistance checks and nozzle changes, idle quality can be restored to an acceptable level.

14. An injection pump change may be necessary if either of the following occurs:

a. If the problem moves from cylinder to cylinder as changes in nozzles are made.

b. If cylinder ohm readings do not change when nozzles are changed.

Note: It is important to always recheck the cylinders at the same RPM. Sometimes the cylinder readings do not indicate that an improvement has been made although the engine may in fact idle better. An injection nozzle with a tip leak can allow more fuel than normal into the cylinder, which will raise the glow plug ohm reading. This will rob fuel from the next injection nozzle (if the injection pump is equipped with a vented rotor) in the firing sequence, and will result in that glow plug having a low ohm reading. If this is encountered, it is advisable to remove and check the injection nozzle with a high reading. If it is leaking, it could be causing the rough idle.

Some glow plugs have been found which do not increase in resistance with heat. If you experience low readings on a glow plug that does not change with injection nozzle change, switch glow plugs between a good and bad cylinder. If the reading of each cylinder is not the same as before the switch, the glow plug cannot be used for rough idle diagnosis, although it will function for starting the car.

4. SMOKE DIAGNOSIS

The smoke emitted from the exhaust system can give valuable clues to the engine's operating condition. The diesel engine emits three different kinds of smoke—black, white, and blue.

Black Smoke

As stated in earlier chapters, black smoke occurs when there is an insufficient amount of air to complete combustion. The fuel that does not unite with the air is heated and changed to soot.

Since the injection pump is incapable of producing an excessively lean or rich mixture, it is not a primary cause for producing smoke. Causes of a black smoke condition are:

- restricted air intake
- air in injection pump
- fuel return circuit restricted
- pump timing advanced
- wrong fuel
- low nozzle-opening pressure
- inadequate fuel supply pump pressure
- restricted exhaust
- low compression

White Smoke

White smoke occurs when the combustion chamber temperature is too low for complete combustion. Because the temperature is so low (about 500°F [260°C] lower than normal), the fuel ignites too late and goes into the exhaust system in a partial or unburned state. Smoke, when cold, is normal but should clear once the engine is warm. Probable causes of white smoke emission;

- low combustion chamber temperature (loss of compression)
- fuel cetane rating too low
- retarded injection pump timing

Blue Smoke

Like the gasoline engine, the diesel emits blue smoke if engine oil enters the combustion chamber. Common causes are:

- faulty crankcase ventilation system
- too high oil level
- worn piston rings
- worn valve seals

5. TURBOCHARGER DIAGNOSIS

Diesel engines with turbochargers (TC) can cause problems not found on naturally aspirated engines. Checking turbocharger performance requires the following:

- appropriate hand tools
- TC boost gauge
- manufacturer's service manual

Diagnose turbocharger performance by following these procedures:

1. Connect TC boost gauge
2. Drive vehicle at specified speed and RPM
3. Read boost pressure

If boost pressure is too high:

1. Boost control line from manifold to wastegate is leaking
2. Defective wastegate

If boost pressure is too low (low-power complaint):

1. Dirty air filter
2. Air leaks in TC system
3. Defective blow-off valve
4. Defective wastegate
5. Defective turbocharger

Be sure to diagnose the cause of a defective turbocharger before replacing it. Common causes may be:

- operator abuse
- overheating
- low oil pressure

SUMMARY

A diesel engine tune-up, consists for the most part, of a series of checks, adjustments, and replacement of filters.

Diesel engine problems are usually the result of contaminated fuel, oil, and dirty air. Among the very last items that cause trouble is the injection pump. The injection pump should only be replaced when it is certain that it is the cause of the problem.

CHAPTER 18 QUESTIONS

1. List the basic services performed when the diesel-powered vehicle is serviced.
2. List in order the procedure you would use for:
   a. no cranking
   b. no starting

    c. rough idle

    d. low power

3. Explain how you would locate a missing cylinder.

4. On what basic principle does the glow plug resistance test rely?

5. List the problems that could cause:

    a. black smoke

    b. white smoke

    c. blue smoke

GLOSSARY

Air/fuel ratio—The amount of air in relation to fuel. Diesel engine ratios can range from 100/1 to 20/1.

Aneroid compensator—A device on the fuel injection pump designed to vary the amount of fuel as atmospheric pressure changes.

Antechamber—On direct injection engines, the chamber in which combustion begins.

Black smoke—Smoke consisting of carbon particulates produced when there is not enough air to unite with the fuel to burn; the fuel is heated and changed to carbon instead.

Blended fuels—A mixture of no. 1D and no. 2D diesel fuels.

Block heater—A device used to warm the engine's coolant, promoting easier starting in cold weather. Usually found in place of one of the core plugs.

Blow-off valve—A device mounted on the intake manifold designed to vent excess pressure to the atmosphere.

Blue smoke—Smoke formed when the engine burns its lubricating (crankcase) oil.

Boost pressure—Air pressure developed by the turbocharger.

Bottom Dead Center (BDC)—The lowest point of piston travel.

British Thermal Unit (BTU)—The amount of heat needed to raise one pound of water one degree Fahrenheit.

Calorie—The amount of heat needed to raise one gram of water one degree Celsius.

Carbonaceous wear—Wear produced by carbon particles.

Carbon monoxide—A poisonous, colorless, odorless, tasteless gas produced by incomplete combustion of the air/fuel mixture.

Cetane rating—A method for measuring the ignition quality of diesel fuel. The higher the cetane number, the lower the ignition delay time.

Charging cycle—The part of injection pump operation in which a metered quantity of fuel is being readied to be pressurized.

Chatter test—One of the injection nozzle tests that measures freedom of needle movement in its bore.

Cloud point (Wax Appearance Point)—The temperature at which wax crystals begin to form in diesel fuel.

Cold start device (CSD)—A manual or automatic device that usually advances injection timing by a few degrees.

Compression-ignition—A means of igniting an air/fuel mixture by heat generated from compressed air. Used in the diesel cycle.

Compression ratio (CR)—The comparison of cylinder volume when the piston is at bottom dead center to cylinder volume when the piston is at top dead center.

Compression stroke—Stroke that occurs when the cylinder is sealed and the piston travels upward, pressurizing and heating the air.

Controlled combustion phase—The third stage in the diesel combustion process in which the greatest amount of power is produced.

Crankcase ventilation system—Ventilation system designed to prevent excess pressure so that too much vacuum does not occur in the crankcase.

Delivery valve—A component of the injection pump designed to end fuel flow to the injection nozzle(s) quickly and help maintain a residual pressure in the injection line.

Diesel cycle—Cycle named after Dr. Diesel which utilizes the four strokes of the Otto cycle but uses the heat of compression for ignition.

Diffusion flame process—The process of combustion in a diesel engine. Combustion starts at more than one point. The rate of combustion is determined by how rapidly the air and fuel mix.

Direct drive—System in which the camshaft is driven by the use of gears rather than a chain or belt.

Direct injection (DI)—System in which fuel is injected into the combustion area directly above the piston.

Discharge cycle—The part of injection pump operation in which a metered quantity of fuel is pressurized and sent to the injection nozzle.

Distributor-type injection pump—Pump that uses one pumping element to pressurize the fuel sent to all of the injection nozzles.

Effective stroke—The distance the plunger moves from the time the fuel is first pressurized to the end of fuel pressurization.

Engine cut-off speed—The maximum engine speed allowed by the governor.

Equalizing stroke—On the VE-type pump, this stroke prevents pressure interference between the fuel supplied to various delivery valves.

Expansion ratio—Opposite of compression ratio. The comparison of cylinder volume when the piston is at top dead center to when the piston is at bottom dead center. It is used to describe the ratio of the expanding gases on the power stroke.

Exhaust Gas Recirculation (EGR) system—This system introduces exhaust gases into the intake air stream, which lowers peak combustion temperatures and the formation of NOx.

Exhaust stroke—The period of time in which the engine cylinder is purged of exhaust gases.

Fixed piston pin—The piston pin is fastened to the piston rather than being free to rotate.

Flash point—The lowest temperature at which the fuel will burn when ignited by an external source.

Fuel injector—A device that takes relatively low pressure fuel and greatly increases the pressure.

Fuel supply pump—An external pump that draws fuel from the tank and sends it to the injection pump.

Full floating piston pin—System in which the piston pin rotates rather than being fixed.

Full fuel position—Position in which the injection pump allows maximum fuel flow to the injection nozzles.

Glow plug—A device used to heat the air in the combustion chamber.

Hybrid chamber—A combustion chamber combining the characteristics of the precombustion and swirl chambers.

Hydrocarbons—Unburned fuel present in exhaust.

Ignition lag (delay time)—The time period from when the fuel is first injected into the cylinder to when it begins to burn.

Indirect drive—System in which the camshaft is driven by a chain or belt rather than by gears.

Indirect injection (IDI)—A type of design in which fuel is first injected into an antechamber. Combustion begins in the antechamber but exits into the main chamber where there is enough air to complete combustion.

Injection nozzle—A component that directs fuel already under high pressure into the cylinder.

Injection timing advance—The injection of fuel into the cylinder earlier than usual to ensure complete combustion as piston speed increases.

In-line injection pump—System that uses a pumping element for each injection nozzle.

Intake stroke—Stroke in which air is drawn into the cylinder primarily by the downward movement of the piston.

Inward opening nozzle—Nozzle in which the nozzle needle moves up into the nozzle allowing fuel to exit.

Keystone ring—A type of compression ring primarily used in the uppermost ring groove because of its resistance to sticking caused by carbon.

Kinetic energy—Energy of motion.

Lag time—The time from when the driver steps on the accelerator pedal to when boost pressure arrives at the cylinder.

Luminosity probe—A device inserted in the glow plug cavity that allows the light of combustion to be seen and sensed by a timing meter.

Match fitted—A part fitted specifically to maintain specified dimensions. For example, pistons are matched and fitted into cylinder bores to maintain the proper tolerances.

Mechanical efficiency—The power expended to keep the engine's internal parts moving.

Mechanical energy—*See* Kinetic energy.

Minimum-maximum governor—A device that controls engine idle speed and maximum engine speed. The speed range is controlled by the driver.

Naturally aspirated—An engine that does not use an external device to force more air into the cylinders. Instead, this engine relies on atmospheric pressure forcing air into the low-pressure area created by the downward movement of the piston.

Non-positive displacement pump—A pump that displaces a varied volume per stroke or revolution. An example would be a turbocharger.

Nozzle—Specifically, the component that contains the nozzle needle.

Nozzle holder—The component that contains and secures the nozzle.

Opening (pop-off) pressure—The point at which injection line pressure overcomes nozzle spring pressure forcing the needle off its seat.

Otto cycle—A cycle developed by Dr. Otto commonly used to power gasoline engines; uses a spark for ignition.

Outward opening nozzles—A nozzle in which the nozzle needle moves out, away from the nozzle allowing fuel to exit.

Overboost—A condition in which too much boost pressure is developed.

Oxides of nitrogen (Nox)—The gases formed when the temperature is maintained high enough to allow nitrogen and oxygen molecules to combine.

Particulates—Particles of soot formed from unburned but heated fuel.

Piston cooling jet—A device that directs a spray of engine oil under the piston, keeping the piston cool.

Positive displacement pump—A pump that displaces the same volume per stroke or revolution. An example would be a Roots-type supercharger.

Pour point—The temperature at which the fuel will no longer flow.

Power stroke—The only stroke that converts the fuel's heat energy into work or power. When the fuel's heat energy is released, cylinder pressure and temperature rise, forcing the piston downward.

Precombustion chamber—Typically, it is characterized by having a long venturi and approximately 30% of the combustion chamber volume.

Pumping element—The unit within the injection pump that pressurizes fuel to the level of injection pressure.

Pumping loss—Energy expended moving gases into and out of an engine.

Rapid combustion phase—The second phase of the diesel combustion process characterized by a rapid rise in pressure.

Return fuel—Excess fuel that is returned to the fuel tank after cooling and lubricating the injection pump and nozzles.

Reverse flow dampening valve—Dampens pressure waves within the injection line caused by the sudden closing of the nozzle needle, preventing secondary injections.

Seat tightness—A test that indicates the condition of the nozzle needle and seat area.

Semi-floating piston pin—System in which the piston pin is fastened to the connecting rod.

Smoke limit—The point at which the addition of more fuel into the cylinder will result in smoke.

Snubber valves—Devices designed to prevent secondary injections.

Specific gravity—A liquid's weight when compared to the same volume of water. Water is assigned a value of one.

Spray pattern—The shape of the fuel spray after exiting the nozzle.

Supercharging—A means of artificially forcing more air into the cylinder that could not be obtained naturally.

Swirl chamber—Typically, a chamber that contains a short venturi and approximately 70% of combustion chamber volume.

Thermal efficiency—The percentage of converting the fuel's heat energy into kinetic (mechanical) energy.

Top dead center (TDC)—The highest point of piston travel.

Vacuum pump—A device providing vacuum to vacuum-assisted controls and servos.

Valve overlap—The number of degrees when both the intake and exhaust valves are open at the same time.

Valve timing—The opening and closing of the intake and exhaust valves at the proper time.

Viscosity—The resistance of a liquid to flow.

Volatility—The ability of a liquid to change to a vapor.

Volumetric efficiency—A percentage of the amount of air drawn into the cylinder on the intake stroke.

Wastegate—A device used on turbochargers to allow exhaust gases to bypass the turbine, limiting boost pressure.

White smoke—Smoked formed when the combustion chamber is cold.

INDEX